T0185700

University of Tehran Science and Humanities Series

Series Editor

University of Tehran Central Acquisitions Office, University of Tehran, Tehran, Iran

The *University of Tehran Science and Humanities Series* seeks to publish a broad portfolio of scientific books, basically aiming at scientists, researchers, students and professionals. The series includes peer-reviewed monographs, edited volumes, textbooks, and conference proceedings. It covers a wide range of scientific disciplines including, but not limited to Humanities, Social Sciences and Natural Sciences.

More information about this series at http://www.springer.com/series/14538

Hooshang Nayebi

Advanced Statistics for Testing Assumed Causal Relationships

Multiple Regression Analysis Path Analysis Logistic Regression Analysis

Hooshang Nayebi
Department of Sociology
University of Tehran
Tehran, Iran

ISSN 2367-1092   ISSN 2367-1106   (electronic)
University of Tehran Science and Humanities Series
ISBN 978-3-030-54756-1   ISBN 978-3-030-54754-7   (eBook)
https://doi.org/10.1007/978-3-030-54754-7

This Springer imprint is published by the registered company Springer Nature Switzerland AG
The registered company address is: Gewerbestrasse 11, 6330 Cham, Switzerland

Preface

The main aim of science is to explain the phenomena. Sciences basically seek to explore cause-and-effect relationships between the phenomena. For example, the social scientist tries to explain why do some people have traditional attitudes and others have modern attitudes? Why is the social participation of women widespread in one place and limited in elsewhere? Why are some people wealthy and some poor? Why do some obey the law and some violate it? And so on.

In an experiment, the researchers vary a phenomenon and examine its effect on another phenomenon, while exactly holding constant the effects of other phenomena. But the causal structure of social phenomena cannot usually be explored in a laboratory setting. However, the multivariate techniques permit to test the hypotheses about cause-and-effect relationships between social phenomena. In the multivariate analysis, researchers do not manipulate a phenomenon to examine its effect. Instead, they examine the effect of an existing phenomenon and use statistical adjustments for holding constant the effects of other phenomena. Also, since social phenomena are usually influenced by a number of factors, the multivariate analysis allows us to examine the simultaneous effects of a series of phenomena on a phenomenon.

Of course, the statistical analysis itself does not explain the causal relationship. The statistical analysis can only indicate that the phenomena change together in a given population. In other words, the statistical relation between phenomena only indicates the covariation between them. To consider a statistical relationship between phenomena as a causal relationship depends on the theoretical reasoning. In fact, it is researcher who theoretically assumes a certain causal relationships between phenomena. Then she or he can use multivariate analysis to test the hypotheses. If it indicates the assumed relationships, it can be used as an empirical verification of the hypotheses.

There are many advanced statistics books that include multivariate analysis. However, some of these books cover a wide range of advanced techniques and are like an encyclopedia (e.g., Field 2013; Foster et al. 2006; Agresti and Finlay 2009; Christensen 2001). As a result, they make the reader frustrating and embarrassing. The categorization of advanced techniques and the assignment of a book to every category will help the reader to better focus and understand it (for a categorization of advanced statistical techniques, see Tabachnick and Fidell 2014). Hence, this book presents a series of the advanced statistics that are mostly used to test the assumed causal relationships.

On the other hand, some books are only about one technique and do not include related techniques. For example, numerous books are about the linear regression, but do not include the logistic regression which complements linear regression (e.g., Allison 1999), or do not include path analysis which is one of the most important techniques for analyzing causal relationships (e.g., Best and Wolf 2015; Draper and Smith 1998; Fox 2016; Montgomery et al. 2012). In fact, there are few books specific to test the assumed causal relationships. This book is for filling this relative vacuum.

Moreover, many of the advanced statistical books mainly deal with complex formulas and calculations. But students do not need them. What is needed is the concept and application of statistical techniques, because statistics techniques are only a tool in the quantitative research. Hence, the emphasis of this book is on the concept and application of multivariate techniques in analyzing causal relationships. Indeed, this book is a nontechnical, nonmathematical book, with a simple language which presents concept and application of statistical techniques by simple examples, because nothing better than simple examples can show their meanings and applications. Therefore, it assumes no previous knowledge of statistics and mathematical background. However, since statistical calculations are done with computer, familiarity with one of the statistical software such as R, SAS, STATA or SPSS is necessary. Here, SPSS which is the most commonly used software is used and its commands of statistical calculations are presented (for this software, e.g., see Nie 1975; Norusis 1989; SPSS Inc 2003 and 2004).

In sum, the advantage of this book is that it concentrates on a particular topic, testing the assumed causal relationships between phenomena and the related statistical techniques: linear regression, path analysis and logistic regression. Here the emphasis is on the conceptions and applications of these techniques by using simple examples. Particularly in the first chapter, it is shown by several examples that linear regression accurately reconstructs the relationships that exist in reality between phenomena.

And most importantly, perhaps for the first time, this book presents all potential effects of each causative phenomenon (cause, or the independent variable as it is common in research texts) on a phenomenon (effect, or the dependent variable). In the statistics books (e.g., Foster et al. 2006) or research methods books (e.g., de Vaus 2002), the discussion of the effects of the independent variables on the dependent variable is generally limited to direct and indirect effects. And the sum of these two effects, which is named as total effect, is implicitly seen as pure effect! Or, the comparison is confined only to compare the standardized regression coefficients and again implicitly they are considered as the indicators of the contributions of the independent variables in the variation of the dependent variable (e.g., Allison 1999)! As Darlington and Hayes point out (2017: 269), the standardized regression coefficient is the most used measure of relative importance or size of the effect of a variable!

But it can be shown by using path analysis that the effects of an independent variable are not limited to direct and indirect effects and may include spurious effect. Most importantly, this book shows that every independent variable has a pure effect on the dependent variable. As a result, the unique contribution of an independent variable to the variation of the dependent variable can be shown. The pure effect of some independent variables is the sum of the direct and indirect effects, but the pure effect of the other independent variables is their effects after removing the effect of the other independent variables on them.

A few books have pointed out the spurious effect in the discussion of path analysis (e.g., Mueller et al. 1977; Blalock 1964, 1971; Foster et al. 2006; Mertler and Vannatta 2010). A few books have also marginally pointed out to the pure effect under another name (semipartial correlation) in the discussion of the linear regression (e.g., Kerlinger and Pedhazu 1973; Tabachnick and Fidel 2014; Darlington and Hayes 2017). But in this book, all potential effects of the independent variables on the dependent variable are summed in a matrix, effect matrix. In this way, the importance of each variable can be compared by direct, indirect, pure effects and the unique contribution to the variation of the dependent variable.

About This Book

This book has been designed as a textbook of applied multivariate analysis for testing the assumed causal relationships for masters-level students, Ph.D. students and researchers of social sciences and behavior sciences, although it can be used in many other disciplines, such as medicine, engineering, the chemical and physical sciences.

It contains three chapters: multiple regression analysis, path analysis and logistic regression analysis. Chapter 1 "Multiple Regression Analysis" describes the multiple linear regression and shows that this technique can accurately reconstruct the causal relationships between variables (phenomena). It also describes the applications of multiple regression which are mainly to test the hypotheses about causal connections and to present the total contribution of the independent variables to the variation of the dependent variable. Moreover, it explains that qualitative variables can be used as binary variables or dummy variables in the linear regression analysis. It also examines the assumptions of multiple linear regression, multi-collinearity and unusual cases. Finally, it presents the command of running multiple regression in SPSS.

Chapter 2 "Path Analysis" deals with the relationships between variables and describes how a causal model about all causal relationships between variables can be examined. It also describes all potential effects of the independent variables on the dependent variable, including spurious and pure effects, and how these causal effects can be obtained. Thus, one can show which variable has the main contribution to the variation of the dependent variable.

The subject of the final chapter "Logistic Regression Analysis" is about a statistical technique to examine the causal effects of the independent variables on the probability of occurrence of an event, which is one of the categories of the dependent variable. Also, it presents how it can specify the pure effect of each of the independent variables on the dependent variable. Finally, it presents the command of running the logistic regression in SPSS.

This textbook is the result of more than sixteen years of teaching the quantitative method (survey) and advanced statistics in the field of sociology and the interaction with students and their comments about statistical techniques. It is partly an adaptation of another author's book in advanced statistics written in Persian (Nayebi 2013).

Tehran, Iran   Hooshang Nayebi

References

Agresti A, Finlay B (2009) Statistical methods for the social sciences, 4th ed. Prentice Hall, Upper Saddle River, New Jersey

Allison PD (1999) Multiple regression: a primer. Sage, Thousand Oaks CA

Best H, Wolf C (eds) (2015) The Sage handbook of regression analysis and causal inference. Sage, London

Blalock HM (1964) Causal inferences in nonexperimental research. University of North Carolina Press, Chapel Hill

Blalock HM (ed) (1971) Causal models in the social sciences. Aldine-Atherton, Chicago

Christensen R (2001) Advanced linear modeling: multivariate time series and spatial data; nonpararmetric regression and response surface maximization, 2nd ed. Springer, New York

Darlington RB, Hayes AF (2017) Regression analysis and linear models, concepts applications and implementation. Guilford Press, New York

de Vaus D (2002) Surveys in social research, 5th ed. Allen & Unwin, Crows Nest NSW

Draper NR, Smith H (1998) Applied regression analysis, 3rd ed. Wiley, New York

Field A (2013) Discovering statistics using, IBM SPSS using, 4th ed. Sage, London

Foster J, Barkus E, Yavorsky C (2006) Understanding and using advanced statistics. Sage, London

Fox J (2016) Applied regression analysis and generalized linear models, 3rd ed. Sage, Thousand Oaks CA

Kerlinger FN, Pedhazur EJ (1973) Multiple regression in behavioral research. New York, Holt Rinehart & Winston

Mertler C, Vannatta R (2010) Advanced and multivariate statistical methods, 6th ed. New York, Routledge

Montgomery DC, Peck EA, Vining GG (2012) Introduction to linear regression analysis, 5th ed. Wiley, Hoboken

Mueller JH, Schuessler KF, Costner HL (1977) Statistical reasoning in sociology, 3rd ed. Houghton Mifflin Company, Boston

Nayebi H (2013) Advanced applied statistics with SPSS. Tehran University Press, Tehran (written in Persian)

Nie NH (1975) SPSS, statistical package for the social science. MacGraw-Hill, New York
Norusis MJ (1989) SPSS 6.1. guide to data analysis. Prentice Hall, Englewood Cliffs NJ
SPSS Inc (2003) SPSS regression models 12.0. SPSS Inc, Chicago
SPSS Inc (2004) SPSS 13.0 base user's guide. SPSS Inc, Chicago
Tabachnick BG, Fidell LS (2014) Using multivariate statistics, 6th ed. Pearson Education, Edinburgh Gate Harlow

The original version of the book was revised: Corrections have been incorporated throughout the book, including the cover. The correction to the book is available at https://doi.org/10.1007/978-3-030-54754-7_4

Contents

This chapter displays how a multiple linear regression can recreate the real relationships between variables (phenomena); how a assumed causal relationship is confirmed; what is the total contribution of the causative variables to the variation of the effect variable; how the effects of variables can be controlled; how the qualitative variables may be used in multiple linear regression; what are the assumptions of multiple linear regression; and how can deal with multi-collinearity and unusual cases.

The *multiple linear regression* is the most widely used multivariate technique in *non-laboratory sciences* such as social sciences for examining the assumed causal relationships between a set of independent variables and a dependent variable. The *dependent variable* is a phenomenon which we seek to explain. And an *independent variable* is a phenomenon which we assume as a *causal factor*, a thing which causes or influences the dependent variable.

Indeed, first we theoretically assume a set of independent variables have simultaneously causal impacts on the dependent variable. Thus, we have a set of *hypotheses* about cause-and-effect relationships between the independent variables and the dependent variable which should be empirically tested. Then, we collect the appropriate data and use multiple linear regression to test our hypotheses.

> Multiple linear regression recreates relationships between a dependent variable and a set of independent variables.

If as we assumed, there are causal relationships between the independent variables and the dependent variable in reality, the multiple linear regression can predict and recreate these relationships as an equation. The linear *regression equation* displays how the dependent variable is related to each of the independent variables.

Here, it is illustrated with a few simple examples.

Example 1.1 In the newly firm A, salary of employees is based on a fixed amount of $2000 and $100 for each additional year of schooling. In addition, male employees are paid $500 more than female employees. In other words, the salary coefficients in the firm A are as follows:

$$\text{Salary} = \$2000 + \$500 \text{ for man} + \$100$$
$$\text{for 1 year of schooling}$$

This firm hired 12 employees. Thus, the employee 1 who is an illiterate woman is paid $2000; the employee 2 who is an illiterate man is paid $2500; the employee 3 who is a woman with 4 years of schooling is paid $2400, and so on (Table 1.1).

Now, suppose a researcher assumed that education (X) has effect on salary (Y): the more education, the more salary. Therefore, she or he interviewed with the employees of the firm and asked their education and salary for testing this hypothesis. Then, these data were entered into a file of statistical software such as SPSS.

The linear regression of salary on education yields the coefficients of the linear regression equation for these two variables (see Fig. 1.1). ◄

H. Nayebi, *Advanced Statistics for Testing Assumed Causal Relationships*, University of Tehran Science and Humanities Series, https://doi.org/10.1007/978-3-030-54754-7_1

Table 1.1 Distribution of gender, education and salary, Example 1.1

No.	Gender[a]	Education (by year)	Salary ($)
1	0	0	2000
2	1	0	2500
3	0	4	2400
4	1	4	2900
5	0	8	2800
6	1	8	3300
7	0	8	2800
8	1	8	3300
9	0	12	3200
10	1	12	3700
11	0	16	3600
12	1	16	4100
Total		96	36,600

[a]0 = female, 1 = male

1.1 Linear Regression Equation

The *linear regression equation* predicts the dependent variable as a function of a set of independent variables:

$$\hat{Y}_i = a + b_1X_1 + b_2X_2 + \cdots + b_jX_j + \cdots \quad (1.1)$$

where $\hat{Y}_i$ is the predicted value of the dependent variable; X_j is jth independent variable; b_j is the regression coefficient of X_j and α is the constant coefficient.

The *linear regression coefficient* (b) of an independent variable (X) represents the change in the dependent variable (Y) for each one unit increase in the independent variable, while controlling for the effects of the other independent variables. The *constant coefficient* (α) of a linear regression equation indicates the average value of the dependent variable when the values of all independent variables are zero.

Thus, the linear regression equation of Example 1.1 is as follows:

$$\hat{Y}_i = a + bX \quad (1.2)$$

The column B in Fig. 1.1 provides the linear regression coefficients for this equation. By replacing these coefficients in the above equation, it will be as follows:

$$\hat{Y}_i = 2250 + 100X \quad (1.3)$$

In this equation, the constant coefficient (α) indicates that the average salary of the illiterate employees (i.e. X = 0) is $2250. The linear regression coefficient of X (education) indicates that Y (salary) increases 100$ for each one year increase in schooling. Therefore, these regression coefficients are almost in accordance with the coefficients of salary in firm A. In other words, it is overt that the regression equation recreated or reconstructed the relationship between salary and education with a high precision.

Here, the linear regression coefficient of X is the same as the coefficient of education in firm A. However, the constant coefficient (α) of the

Coefficients[a]

Model		Unstandardized Coefficients		Standardized Coefficients		
		B	Std. Error	Beta	t	Sig.
1	(Constant)	2250.000	145.774		15.435	.000
	X Education	100.000	15.309	.900	6.532	.000

a. Dependent Variable: Y Salary

Fig. 1.1 Coefficients of the regression of salary on education, Example 1.1. This output is produced by Linear Regression Command in SPSS (this command is described at the end of this chapter, Sect. 1.9)

regression equation is slightly different from the fixed salary of $2000 in firm A. The reason of this difference is that the fixed amount of salary in firm A ($2000) is the salary of employees who are female and illiterate, while the constant coefficient in the regression equation ($2250) is the average salary of employees who are illiterate (X = 0). By calculating the average salary of illiterate employees,[1] we see that it is the same as the constant coefficient of the regression equation. Therefore, the constant coefficient of the above regression equation accurately recreated the reality of salary in firm A, too.

Example 1.2 Now, suppose another researcher based on the assumption that gender affects salary (on average, men earn more than women), interviewed with the employees of firm A and asked their salary and noted their gender. Then, these data were entered into the SPSS file. ◄

The regression of salary on gender yields the coefficients of the linear regression equation for these two variables (Fig. 1.2). Thus, the linear regression equation of these two variables is:

$$\hat{Y}_i = 2800 + 500X \qquad (1.4)$$

This regression equation is almost in accordance with the coefficients of salary in firm A, too. In the above equation, the constant coefficient (α) is $2800; this is the average salary of the female employees (X = 0). The regression coefficient of X (gender) is $500, which means men are averagely paid $500 more than women.

> Linear regression coefficient indicates the change in the dependent variable for one unit change in the independent variable.

Again, it is obvious that the regression equation recreated the relationship between salary and gender with a high precision. Here, the coefficient of X is the same as the coefficient of gender

in firm A. However, as Example 1, the constant coefficient of the regression equation is slightly different from the fixed amount of $2000 in firm A. Because, as mentioned above, the fixed amount of salary in firm A ($2000) is the salary of employees who are female and illiterate, while the constant coefficient (α) in the above regression equation ($2800) is the average salary of employees who are women. By calculating the average salary of female employees, we see it is the same as the constant coefficient of regression equation.[2] Thus, again the constant coefficient (α) of regression equation accurately recreated the reality of salary in firm A.

Example 1.3 This time, suppose another researcher based on the assumption that both gender and education affect salary (men averagely earn more than women and the more education, the more salary), interviewed with the employees of the firm A and asked their salary and education and noted their gender. ◄

The output of the linear regression salary on gender (X1) and education (X2) is presented in Fig. 1.3. Thus, the multiple linear regression equation of these three variables is as follows:

$$\hat{Y}_i = 2000 + 500X_1 + 100X_2 \qquad (1.5)$$

In this example, the multiple linear regression equation quite accurately recreates all coefficients of salary in firm A. The constant coefficient (α) in the linear regression equation shows that the average salary of employees who are women (X1 = 0) and illiterate (X2 = 0) is $2000. The regression coefficient of X1 (gender) means that, on average, male employees (X1 = 1) are paid $500 more than female employees while controlling for education. The regression coefficient of X2 (education) means that salary increases $100 for each one year increase in schooling while controlling for gender.

Here, unlike the two previous equations, the constant coefficient (α) of regression equation is

---

[1]Table 1.1 shows two employees are illiterate. Thus, their average salary is 2250$:$\frac{2000+2500}{2} = 2250$.

[2]Table 1.1 shows six employees are woman. Thus, their average salary is 2800$:$\frac{2000+2400+2800+2800+3200+3600}{6}$ $= \frac{16800}{6} = 2800$

Coefficients[a]

Model		Unstandardized Coefficients		Standardized Coefficients		
		B	Std. Error	Beta	t	Sig.
1	(Constant)	2800.000	230.940		12.124	.000
	X Gender	500.000	326.599	.436	1.531	.157

a. Dependent Variable: Y Salary

Fig. 1.2 Coefficients of the regression of salary on gender, Example 1.2

Coefficients[a]

Model		Unstandardized Coefficients		Standardized Coefficients		
		B	Std. Error	Beta	t	Sig.
1	(Constant)	2000.000	.000		.	.
	X1 Gender	500.000	.000	.436	.	.
	X2 Education	100.000	.000	.900	.	.

a. Dependent Variable: Y Salary

Fig. 1.3 Coefficients of the regression of salary on gender and education, Example 1.3

the same as the fixed salary of employees in firm A, because both represent average salary of employees who are woman and illiterate.

Example 1.3 suggests when the multiple linear regression equation involves all independent variables affected the dependent variable, its constant coefficient (α) is the same as the constant in reality.

1.1.1 Direction of Relationship

In addition to show the relationship between the dependent variable and an independent variable, the regression coefficient indicates the *direction of relationship*. If the coefficient is positive, the relationship between two variables is direct (*ascending*), as shown in Fig. 1.4a. A *positive coefficient* indicates the value of the dependent variable is increased by increasing the value of the independent variable.

If it is negative, the relationship is inverse (*descending*), as Fig. 1.4b. A *negative coefficient*

indicates the value of the dependent variable is decreased by increasing the value of the independent variable.

1.2 Uncorrelated Independent Variables

The *uncorrelated independent variables* are variables that have not relationship with each other. In multiple linear regression analysis, the entering or removing uncorrelated variables don't change their coefficients. For example, in firm A, the *correlation*[3] between gender and education is zero (Fig. 1.5); it means that the education level of women and men are not different. Thus, they are uncorrelated variables. That's why the coefficient of education in Eq. 1.3

---

[3]The correlation, is an index of the association between two quantitative variables, and varies from 0 to 1 (or −1); 0 indicates the lack of association and 1 (or −1) indicates the perfect association.

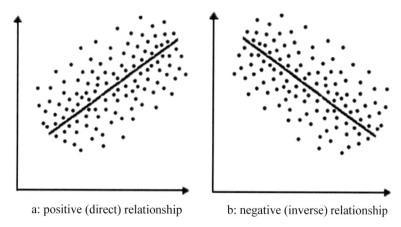

a: positive (direct) relationship   b: negative (inverse) relationship

Fig. 1.4 Direction of the relationship between two variables in a linear regression

Correlations

		X1 Gender	X2 Education
X1 Gender	Pearson Correlation	1	.000
	Sig. (2-tailed)		1.000
	N	12	12
X2 Education	Pearson Correlation	.000	1
	Sig. (2-tailed)	1.000	
	N	12	12

Fig. 1.5 Correlations of gender and education, Example 1.3. This output is produced by Correlation Command (see Sect. 1.9)

(in which education is only independent variable) and Eq. 1.5 (with two independent variables education and gender) is the same.

independent variables change in different regression equations.

Here, it is illustrated by following example.

1.3 Correlated Independent Variables

Correlated independent variables are variables that are connected with each other. In multiple linear regression analysis, the entering or removing correlated variables change their coefficients. Since the coefficient of each independent variable is calculated while controlling for the other independent variables (see Sect. 1.5), the effect of the removed or added correlated variables make the coefficients of the

In multiple linear regression analysis of the different subsets of independent variables, the coefficients of uncorrelated variables don't change.

Example 1.4 Let's assume that the salary coefficients in firm B are as follows:

$$\text{Salary} = \$2000 + \$500 \text{ for man} + \$100$$

$$\text{for 1 year of schooling}$$

$$+ \$140 \text{ for 1 year of service}$$

Table 1.2 Distribution of gender, education, service and salary, Example 1.4

No.	X1 Gender[a]	X2 Education	X3 Service	Y Salary
1	0	0	0	2000
2	1	4	0	2900
3	0	0	0	2000
4	1	4	1	3040
5	0	0	1	2140
6	1	4	2	3180
7	0	0	1	2140
8	1	8	0	3300
9	0	0	2	2280
10	1	8	1	3440
11	0	0	2	2280
12	1	8	2	3580
13	0	4	0	2400
14	1	12	0	3700
15	0	4	1	2540
16	1	12	1	3840
17	0	4	2	2680
18	1	12	2	3980
19	0	8	0	2800
20	1	16	0	4100
21	0	8	1	2940
22	1	16	1	4240
23	0	8	2	3080
24	1	16	2	4380
Total		156	24	72,960

[a]0 = female, 1 = male

The gender, education, service (the duration of service) and salary of employees in this firm are presented in Table 1.2. ◀

Now, suppose we have interviewed with the employees of the firm B and asked their education, the duration of service and monthly salary, and noted their gender and entered these data into the SPSS file. Here, if we only consider the impact of education on salary, the coefficients of the regression of salary on education which is presented in Fig. 1.6, yields the following equation:

$$\hat{Y}_i = 2185.0 + 131.5X \quad (1.6)$$

In the above equation, the value of the coefficient of education slightly differs from the value of the coefficient of education in the salary in firm B. The reason for this difference is that the education is a correlated variable (see Fig. 1.7) and the lack of its correlated independent variable (gender) in this regression analysis has made some of its effect reflected in the coefficient of education.

However, as Fig. 1.8 shows the regression of salary (Y) on gender (X1) and education (X2) yields this regression equation:

$$\hat{Y}_i = 2140 + 500X_1 + 100X_2 \quad (1.7)$$

In this equation, the value of the coefficient of education (X2) is the same as it in the monthly salary in firm B. Here, its correlated independent variable (gender) is entered the regression analysis and makes the coefficient of education

Coefficients[a]

Model		Unstandardized Coefficients		Standardized Coefficients		
		B	Std. Error	Beta	t	Sig.
1	(Constant)	2185.045	74.169		29.460	.000
	X2 Education	131.532	8.865	.953	14.837	.000

a. Dependent Variable: Y Salary

Fig. 1.6 Coefficients of the regression of salary on education, Example 1.4

Correlations

		X1 Gender	X2 Education	X3 Service
X1 Gender	Pearson Correlation	1	.664[**]	.000
	Sig. (2-tailed)		.000	1.000
	N	24	24	24
X2 Education	Pearson Correlation	.664[**]	1	.000
	Sig. (2-tailed)	.000		1.000
	N	24	24	24
X3 Service	Pearson Correlation	.000	.000	1
	Sig. (2-tailed)	1.000	1.000	
	N	24	24	24

[**]. Correlation is significant at the 0.01 level (2-tailed).

Fig. 1.7 Correlations of gender and education and the service, Example 1.4

Coefficients[a]

Model		Unstandardized Coefficients		Standardized Coefficients		
		B	Std. Error	Beta	t	Sig.
1	(Constant)	2140.000	40.072		53.404	.000
	X1 Gender	500.000	66.753	.344	7.490	.000
	X2 Education	100.000	6.336	.725	15.783	.000

a. Dependent Variable: Y Salary

Fig. 1.8 Coefficients of the regression of salary on gender and education, Example 1.4

change. And, because gender is its only correlated variable, its coefficient is equal to the value of the coefficient of education in the monthly salary.

Finally, the regression of the dependent variable on all independent variables yields the following regression equation (Fig. 1.9):

$$\hat{Y}_i = 2000 + 500X_1 + 100X_2 + 140X_3 \quad (1.8)$$

In multiple linear regression analysis of the different subsets of independent variables, the coefficients of correlated variables change.

In this equation, the addition of third independent variable (the duration of service) doesn't change the coefficient of education, because it is an uncorrelated variable. Also, this regression equation quite accurately recreates the coefficients of salary in firm B, since it includes all independent variables which affect the dependent variable. Thus, where the multiple linear regression analysis includes all independent variables, the regression equation will quite accurately recreate the relationships between the independent variables and the dependent variable.

Coefficients[a]

Model		Unstandardized Coefficients		Standardized Coefficients		
		B	Std. Error	Beta	t	Sig.
1	(Constant)	2000.000	.000		.	.
	X1 Gender	500.000	.000	.344	.	.
	X2 Education	100.000	.000	.725	.	.
	X3 Service	140.000	.000	.157	.	.

a. Dependent Variable: Y Salary

Fig. 1.9 Coefficients of the regression of salary on gender, education and service, Example 1.4

1.4 Application of Linear Regression

The linear regression equation is essentially to predict the values of the dependent variable from the independent variables. Therefore, it can be used to predict the dependent variable for each case, such as to predict the job performance of an individual based on a set of factors (independent variables).

> The regression equation is a completely accurate recreating of the relationship between independent variables and a dependent variable when it includes all effective independent variables.

However, in social sciences the linear regression is typically used to *test hypotheses* about the effects of a set of independent variables on the dependent variable. Then, if the hypotheses are confirmed, the total contribution of the independent variables to the variation of the dependent variable can be determined. The linear regression is used in other statistical techniques such as the path analysis which is described in Chap. 2.

1.4.1 Test of Hypotheses

The linear regression of the dependent variable on an independent variable or a set of independent variables is named a *liner regression model*. The relationships between variables of a liner regression model are based on a theoretical reasoning which assumes the independent variables are factors that simultaneously affect the dependent variable. Therefore, the first and most important stage in the linear regression analysis is to show whether the assumed model fits the facts (data) in a population (or a probability sample of it).

A *fitted linear regression model*, which indicates that all hypotheses are confirmed, is one that:

1. All regression coefficients are nonzero,
2. All observed relationships of the independent variables with the dependent variable correspond to the theoretical expectations,
3. If data are collected from a probability sample, the statistical significance levels of all regression coefficients are at least less than 0.05.

Thus, the second regression model in Example 1.4 (the regression of salary on gender and education of all employees) is a fitted regression model, since its coefficients are not zero (Fig. 1.8) and all relations are in accordance with the theoretical expectations (men averagely earn more than women and the more education, the more salary).

Moreover, if that data were collected from a probability sample, the significance level of each of the coefficients should be less than 0.05.

But what is the significance level of a regression coefficient?

Significance Level

The *significance level* of the regression coefficient of an independent variable is the probability of occurrence of such a coefficient under the null hypothesis that there is no relationship between dependent variable and that independent variable in population. Indeed, the *null hypothesis* assumes there is no relationship between two given variables in the population from which the sample is derived. Consequently, if the null hypothesis is true, the observed coefficient (connection) between variables is due to sampling error.

However, where the probability of occurrence of such a coefficient which is observed in the sample is weak (typically less than 0.05), then the null hypothesis is rejected. Therefore, it can be concluded with a given probability (usually of 95%), that the dependent variable is related to that independent variable in the population. In other words, what is observed in the sample is more or less in the population. Thus, when we say a regression coefficient is significant it means that the value of the regression coefficient isn't due to sampling error and it can be generalized to the population.

> In a fitted regression model, all regression coefficients are nonzero, all relations are according to the theoretical expectations and in the case of sampling the significance levels of all coefficients are less than 0.05.

The following example illustrates a fitted regression model by using true data.

Example 1.5 Let's assume that the occupational prestige as social position of individual is affected by gender and education, i.e. men occupy prestigious occupations more than women and individuals with higher education have more access to higher prestigious occupations. Here, we test these hypotheses by using the data of the General Social Survey 1972[4] which is a probability sample of US people.[5]

The coefficients of the regression of occupational prestige on gender and education (Fig. 1.10) indicates that the regression model of Example 1.5 is a fitted regression model and both hypotheses are confirmed, because: (1) its coefficients are not zero; (2) all relations are according to our theoretical expectations (men occupy prestigious occupations more than women and individuals with higher education occupy higher prestigious occupations); and (3) in this probability sample, the significance level of both linear regression coefficients are less than 0.05 (Fig. 1.10, column Sig.).[6] ◄

1.4.2   Total Contribution

The contribution of the independent variables to the variation of the dependent variable is presented by a measure called coefficient of determination, which is also called *measure of goodness of fit*. The *coefficient of determination*, denoted by R^2 (R squared) indicates the proportion of the variation in the dependent variable that is explained by the independent variables in a fitted linear regression model:

$$R^2 = \frac{Explained\ Variation}{Total\ Variation} \qquad (1.9)$$

For example, in the fitted linear regression model of Example 1.5, the coefficient of determination indicates that about 31 percent of the variation (variance) of occupational prestige is explained by gender and education (Fig. 1.11). In other words, 31% of the difference between occupational prestiges of individuals is due to their gender and education. The remaining difference is due to other factors that remain unknown or are not included in the regression model.

The values of the coefficient of determination range from 0 to 1. The coefficient of determination of zero indicates the lack of relationship

---

[4]http://www.gss.norc.org/get-the-data/spss.

[5]In this file, we used PRESTIGE (R's occupational prestige score) and EDUC (Highest year of school completed). But, we turned SEX (Respondents sex) into

a different variable, named SEXR as female = 0 and male = 1.

[6]Do you think this regression model is a fitted one nearly half a century later? You can test your hypotheses by the data of the General Social Survey 2018 (see Exercise 1.9).

Coefficients[a]

Model		Unstandardized Coefficients		Standardized Coefficients		
		B	Std. Error	Beta	t	Sig.
1	(Constant)	12.798	1.071		11.950	.000
	SEXR Respondents sex	1.904	.594	.070	3.202	.001
	EDUC Highest year of school completed	2.155	.086	.550	25.101	.000

a. Dependent Variable: PRESTIGE R's occupational prestige score(1970)

Fig. 1.10 Coefficients of the regression of occupational prestige on sex and education, Example 1.5

Model Summary

Model	R	R Square	Adjusted R Square	Std. Error of the Estimate
1	.556[a]	.309	.308	11.258

a. Predictors: (Constant), EDUC Highest year of school completed, SEXR Respondents sex

Fig. 1.11 Model Summary of the regression of occupational prestige on sex and education, Example 1.5

between the dependent variable and the independent variable. But, where the linear regression equation includes all independent variables which affect the dependent variable, the coefficient of determination will reach its maximum value of 1, which means the total variation of the dependent is explained by the independent variables.

> The coefficient of determination indicates the proportion of the variation of the dependent variable explained by the independent variables.

1.5 Controlling for Independent Variables

The linear regression coefficient of each independent variable is calculated by controlling for the other independent variables. But what does control mean? The *controlling* in linear regression means to remove the effects of the other independent variables from the independent variable X and the dependent variable Y. This removal of the effect is applied in the calculation formula of the regression coefficient of each independent variable. However, the control of effects can be illustrated in a simple way.

The differences between *observed values* (Y_j) of the dependent variable and its *predicted values* ($\hat{Y}_i$) which obtained by the regression equation are the *residuals* of that variable after remove the effects of the independent variables from it. Indeed, the residuals in the regression of a variable on a set of variables remain as a new variable which is free of the effects of those variables.

Thus, if we save the residuals of the regression of an independent variable on the other independent variables as a new variable, and also the residuals of the regression of the dependent variable on those independent variables as another variable, the regression of the latter on the former is the same coefficient of the independent variable in the multiple regression equation.

For example, the linear regression coefficient of X2 (education) in the regression of salary on gender, education and service (Eq. 1.8, Example 1.4) is 100. Now, for calculating the coefficient of X2 by using residuals, we save residuals of the regression of X2 on two other independent variables (X1 and X3) as RESX2 (Table 1.3, column 9) and residuals of the regression of Y on those two independent variables as RESY (Table 1.3, column 7).[7] Then, the regression of RESY on RESX2 results this equation (see Fig. 1.12):

$$RESY = 0.000 + 100RESX2 \qquad (1.10)$$

This coefficient in a simple regression of one independent variable (after removed the effects of the other independent variables) is the same as its coefficient in a regression with other independent variables in Eq. 1.8.

1.6 Linear Regression Methods

In social sciences, the multiple linear regression analysis is essentially for testing hypotheses and the appropriate regression method for it is enter method.[8] The enter method is a single-step method. In this method, all variables enter the regression analysis at once and all independent variables and the dependent variable constitute one linear regression model which is to be tested.

In enter method, the linear regression model will be either confirmed or rejected. As mentioned previously, a confirmed (fitted) regression model is one that all regression coefficients are nonzero and in the case of probability sample, all coefficients are significant at level less than 0.05. Also, the model is consistent with theoretical expectations of researcher, i.e. all of hypotheses are confirmed. But if even one of hypothesis isn't confirmed, the linear regression model will totally be rejected.

Thus, the enter method is the appropriate multiple linear regression method for testing a given model of the relationships between a set of variables. Therefore, it can be said the enter method is the main method of linear regression in social sciences.

However, there are other methods which are used to explore statistical relations between a set of variables and a dependent variable, such as forward method, backward method and stepwise method. All of these methods are *multi-stage methods* or *hierarchical methods* of linear regression: the independent variables depending on their contribution to the dependent variable are included in or excluded from the regression analysis step by step. Every step of including or excluding independent variables is identified as a linear regression model.

Anyway, these multi-stage methods which essentially are for exploring statistical relations aren't focus of this book which focuses on testing the causal relations.

> The enter method is the appropriate regression method for testing a given set of hypotheses.

---

[7]The regression of Y on X1 and X3 yields this equation: $\hat{Y}_i = 2300 + 1200X1 + 140X3$. So, the predicted value of case 1 will be: $\hat{Y}_1 = 2300 + 1200(0) + 140(0) = 2300$; the predicted value of case 2 will be $\hat{Y}_2 = 2300 + 1200(1) + 140(0) = 3500$ and so forth (Table 1.3, column 6). And, the residual of case 1 will be: $2000 - 2300 = -300$; the residual of case 2 will be: $2900 - 3500 = -600$ and so on (Table 1.3, column 7). Also, the regression of X2 on X1 and X3 yields this equation: $\hat{X}2_i = 3 + 7X1 + 0X3$. So, the predicted value of case 1 will be: $\hat{X}2_1 = 3 + 7(0) + 0 = 3$; the predicted value of case 2 will be $\hat{X}2_1 = 3 + 7(1) + 0 = 10$ and so forth (Table 1.3, column 8). And, the residual of case 1 will be: $0 - 3 = -3$; the residual of case 2 will be: $4 - 10 = -6$ and so on (Table 1.3, column 9).

[8]The default method of linear regression in SPSS is the enter method, too.

1.7 Qualitative Variables in Linear Regression

Although the linear regression is basically specific to the quantitative (continuous) variables, the *qualitative variables* which are

Table 1.3 Distribution of gender, education, service, salary and predicted value and residual, Example 1.4

No.	X1	X2	X3	Y_i	$\hat{Y}_i$	RESY	$\hat{X}2$	RESX2
(1)	(2)	(3)	(4)	(5)	(6)	(7)	(8)	(9)
1	0	0	0	2000	2300	−300	3	−3
2	1	4	0	2900	3500	−600	10	−6
3	0	0	0	2000	2300	−300	3	−3
4	1	4	1	3040	3640	−600	10	−6
5	0	0	1	2140	2440	−300	3	−3
6	1	4	2	3180	3780	−600	10	−6
7	0	0	1	2140	2440	−300	3	−3
8	1	8	0	3300	3500	−200	10	−2
9	0	0	2	2280	2580	−300	3	−3
10	1	8	1	3440	3640	−200	10	−2
11	0	0	2	2280	2580	−300	3	−3
12	1	8	2	3580	3780	−200	10	−2
13	0	4	0	2400	2300	100	3	1
14	1	12	0	3700	3500	200	10	2
15	0	4	1	2540	2440	100	3	1
16	1	12	1	3840	3640	200	10	2
17	0	4	2	2680	2580	100	3	1
18	1	12	2	3980	3780	200	10	2
19	0	8	0	2800	2300	500	3	5
20	1	16	0	4100	3500	600	10	6
21	0	8	1	2940	2440	500	3	5
22	1	16	1	4240	3640	600	10	6
23	0	8	2	3080	2580	500	3	5
24	1	16	2	4380	3780	600	10	6
Total		156	24	72,960	72,960	0	156	0

Coefficients[a]

Model		Unstandardized Coefficients		Standardized Coefficients		
		B	Std. Error	Beta	t	Sig.
1	(Constant)	1.850E-15	.000		.	.
	RESX2 Unstandardized Residual	100.000	.000	1.000	.	.

a. Dependent Variable: RESY Unstandardized Residual

Fig. 1.12 Coefficients of the regression of RESY on RESX2

categorical variables or ordinal variables may also be used as binary variables or dummy variables in the linear regression analysis.

1.7.1   Binary Variable

Binary variable also called dichotomous variable or binomial variable is a variable with two categories, such as the participation in an election (voted or not voted). A binary variable is sufficient to be coded 0 and 1 in order to be used in the linear regression analysis such as gender that has been coded as female = 0 and male = 1 in previous examples.

The regression coefficient of binary variable represents how much the value of the dependent variable on the category with code 1 of the binary variable averagely differs from the value of it on the other category (coded 0) while controlling for the effects of the other independent variables.

Although, to give code 0 or 1 to this or that category doesn't matter, it is better to give the code 0 to the category with lower value in order to facilitate interpretation of the regression coefficients of a binary variable. In this case, the linear regression coefficient of binary independent variable indicates how much the value of the dependent variable on the category with code 1 is more than its value on the other category.

Multi-categories Variables

A multi-categories variable which is a qualitative variable with more than two categories may be turned into binary variable and used in linear regression analysis. In this way, it involves a category which is to be compared with a new category which includes all of other categories.

Here, it is illustrated by following example.

> A binary variable or dichotomous variable or binomial variable is a variable with two categories.

Example 1.6 Table 1.4 is the distribution of race (X1), education (X2) and monthly income (Y) of a probability sample of people in a town.

Table 1.4 Distribution of race, education, recoded race and income, Example 1.6

No.	X1[a] Race	X2 Education	X1R[b] Recoded race	Y Income
1	1	0	0	2000
2	1	2	0	2200
3	2	0	0	2200
4	2	2	0	2400
5	2	4	0	2600
6	3	2	1	2700
7	3	4	1	2900
8	3	4	1	2900
9	3	6	1	3100
10	3	8	1	3300
Total		32		26,300

[a]1 = Race A, 2 = Race B and 3 = Race C
[b]1 = Race C and 0 = Race Other

The race which has three categories (A, B and C) can be turned into a binary variable and used in the linear regression analysis. For this purpose, it can be recoded into a new variable X1R (Recoded Race) with two races: Race C and Race Other. Race C (coded 1) is the same race and Race Other (coded 0) involves races A and B (Table 1.4). ◄

Figure 1.13 presents the coefficients of the regression of the dependent variable Y on the independent variables X1R and X2 which yields this equation:

$$\hat{Y}_i = 2108.0 + 356.0X1R + 107.5X2 \quad (1.11)$$

Here, the constant coefficient indicates that the average monthly income of individuals who are illiterate (education = 0) and from the races other than race C (X1R = 0) is $2108. The coefficient of X1R indicates that on average, the monthly income of people of race C (code = 1) is $356 more than people of other races (code = 0) while controlling for education. And the coefficient of X2 means that the monthly income increases $107.5 for one year increase in education while controlling for the effect of X1R.

Coefficients[a]

Model		Unstandardized Coefficients		Standardized Coefficients		
		B	Std. Error	Beta	t	Sig.
1	(Constant)	2108.000	42.988		49.037	.000
	X1R Recoded Race	356.000	68.935	.439	5.164	.001
	X2 Education	107.500	14.361	.637	7.485	.000

a. Dependent Variable: Y Salary

Fig. 1.13 Coefficients of the regression of Y on X2 and X1R, Example 1.6

1.7.2 Dummy Variable

The qualitative variables with more than two categories may be turned into a set of binary variables called dummy variables and applied in the linear regression analysis. For this purpose, except one category of the original variable, each of other categories is turned into a new binary variable. A category which is excluded from being turned into a binary variable is as base category for comparing dummy variables with it. Thus, the number of dummy variables will be equal to the number of categories of original variable minus one. In other words, if a qualitative variable holds k categories, the 1-k dummy variables are constructed.

> The regression coefficient of a binary independent variable shows the change in the dependent variable from one of its category to other category.

Every *dummy variable* indicates the presence of a given category of the original variable (coded 1) and the absence of other categories (coded 0). The linear regression coefficient of dummy variable represents that on average how much the value of the dependent variable on the category with code 1 of that dummy variable differs from the value of it on the base category while controlling for the effects of the other independent variables.

Although, as mentioned previously about binary variables, to select this or that category as the base one doesn't matter, it is better to select one that the values of the dependent variable in it are averagely lower than others in order to facilitate the interpretation of the regression coefficients of dummy variables.

Thus, in above example, we can turn the race (X1) into two dummy variables in order to compare the income of all three races. For this purpose, we select Race A as the base category, because the mean of income of its members is lower than the means of other categories (see Fig. 1.14). Then, we construct a dummy variable of XB with a category involves members of Race B (coded 1) and another category involves members of other races (coded 0). Also, we construct another dummy variable of XC with a category involves members of Race C (coded 1) and another category involves members of other races (coded 0).[9] These two dummy variables are presented in Table 1.5.

Figure 1.15 presents the coefficients of the regression of Y on the independent variables of X2 (education), XB (Race B) and XC (Race C) which yields this equation:

$$\hat{Y}_i = 2000 + 100X2 + 200XB + 500XC \quad (1.12)$$

> A dummy variable indicates the presence of a given category of the original variable and the absence of its other categories.

___
[9]The dummy variables can be obtained by Dummy Variable Command which is presented at Sect. 1.9.

Report

Y Salary

X1 Race	Mean	N	Std. Deviation
1 Race A	2100.00	2	141.421
2 Race B	2400.00	3	200.000
3 Race C	2980.00	5	228.035
Total	2630.00	10	427.005

Fig. 1.14 Report of mean of Y by X1, Example 1.6. This output is produced by Compare Mean Command which is presented at the end of the chapter, Sect. 1.9

Table 1.5 Distribution of race, education, race B, race C and income

No.	X1[a] Race	X2 Education	XB[b] Race B	XC[c] Race C	Y Income
1	1	0	0	0	2000
2	1	2	0	0	2200
3	2	0	1	0	2200
4	2	2	1	0	2400
5	2	4	1	0	2600
6	3	2	0	1	2700
7	3	4	0	1	2900
8	3	4	0	1	2900
9	3	6	0	1	3100
10	3	8	0	1	3300
Total		32			26,300

[a] 1 = Race A, 2 = Race B and 3 = Race C
[b] 1 = Race B and 0 = other races
[c] 1 = Race C and 0 = other races

Here, the constant coefficient of linear regression shows that the average monthly income of individuals who are not from races B (XB = 0) and C (XC = 0) and are illiterate (X2 = 0) is $2000. The coefficient of X2 means that the monthly income increases $100 for one year increase on education while controlling for the effect of XB and XC. The coefficient of XB indicates that the average monthly income of members of race B (code = 1) is $200 more than people of race A (base category) while controlling for education. The coefficient of XC indicates that the average monthly income of members of race C (code = 1) is $500 more than the members of race A while controlling for education.

This example used a hypothetical data, but the next example uses a true data.

Regression coefficient of a dummy variable indicates how much the value of the dependent variable in the category of code 1 of that dummy variable is different from its value in the base category.

Example 1.7 Let's assume race and education affect individual occupational prestige: the majority occupies prestigious occupations more than the minority. Also, individuals with higher education have more access to higher prestigious occupations. Here, we test these hypotheses by using the data of the General Social Survey 2018[10] which is a probability sample of US people. In this file, race has three categories: white, black and other. Therefore, we turn it into two dummy variables: RACE1 (WHITE) and RACE3 (OTHER). We take black as the base category for comparing dummy variables with it. ◀

Figure 1.16 presents the coefficients of the regression of the occupational prestige (PRESTG10) on education (EDUC) and RACE1

_____
[10]http://www.gss.norc.org/get-the-data/spss. In this file, the occupational prestige is PRESTG10 (R's occupational prestige score), education is EDUC (Highest year of school completed) and race is RACE (Race of respondent).

Coefficients^a

Model		Unstandardized Coefficients		Standardized Coefficients		
		B	Std. Error	Beta	t	Sig.
1	(Constant)	2000.000	.000		709244518.200	.000
	X2 Education	100.000	.000	.592	143611069.400	.000
	XB Race B	200.000	.000	.226	55620328.000	.000
	XC Race C	500.000	.000	.617	119675891.200	.000

a. Dependent Variable: Y Salary

Fig. 1.15 Coefficients of the regression of Y on the independent variables of X2, XB and XC

Coefficients^a

Model		Unstandardized Coefficients		Standardized Coefficients		
		B	Std. Error	Beta	t	Sig.
1	(Constant)	11.685	1.314		8.893	.000
	EDUC Highest year of school completed	2.189	.086	.474	25.436	.000
	RACE1 RACE=WHITE	3.397	.696	.111	4.883	.000
	RACE3 RACE=OTHER	3.028	.986	.069	3.070	.002

a. Dependent Variable: PRESTG10 R's occupational prestige score (2010)

Fig. 1.16 Coefficients of the regression of PRESTG10 on EDUC, RACE1 and RACE3, Example 1.7

(WHITE) and RACE3 (OTHER). This table indicates that our hypotheses are confirmed, since all coefficients are nonzero and all relations are in accordance with our theoretical expectations: white (the majority) occupies prestigious occupations more than the minority; also individuals with higher education have more access to higher prestigious occupations. Also, the significance levels of all three regression coefficients are less than 0.05. Thus, this regression model is a fitted regression model.[11] The regression equation of this fitted model is:

$$PRESTAG10 = 11.685 + 2.189EDUC$$
$$+ 3.397RACE1 + 3.028RACE3$$
$$(1.13)$$

Here, the constant coefficient shows that on average the occupational prestige (PRESTG10) score of individuals who are not white and other races and are illiterate (EDUC = 0) is 11.685. The coefficient of EDUC means that the occupational prestige increases 2.189 score for one year increase on schooling while controlling for the effect of RACE 1 (white) and RACE 3 (other races). The coefficient of RACE1 indicates that on average, the occupational prestige of white is 3.397 score more than black (base category) while controlling for education. And the coefficient of RACE3 indicates that on average, the occupational prestige of other races is 3.028

---

[11]Do you think these relationships were also true nearly half a century ago? You can test these hypotheses by the data of the General Social Survey 1972 (see Exercises 1.14–1.16).

score more than black while controlling for education.

1.8 Linear Regression Considerations

The multiple linear regression is based on assumptions that the serious violations of them may distort linear regression analysis. Multicollinearity which indicates the very strong correlation between the independent variables is another important problem. Also, unusual or far-flung cases may also reduce fitness of the regression model. In this section, these three issues are briefly described.

1.8.1 Linear Regression Assumptions

The *assumptions* of linear regression are:
1. The *measurement level* of the independent variables can be either quantitative (continuous and interval) or binary (dichotomous) but of the dependent variable must be only quantitative.
2. The relationship between the dependent variable and each quantitative independent variable is linear and additive.
3. The observations are independent of each other.
4. The distribution of the dependent variable for each value of the independent variable is normal.
5. The variance of the dependent variable is the same for all values of the independent variables.

In general, one should not be very concerned about assumptions, in particular where the *sample size* is large. Indeed, linear regression is a *robust technique*, although severe violations of its assumptions may alter the results of regression (see Kerlinger and Pedhazur 1973).

However, it is necessary to examine these assumptions in order to ensure that they are not severely violated.

Measurement Level of Variables
The first assumption of linear regression is that all variable except the binary and dummy independent variables are quantitative. The *quantitative variable*, also named *continuous variable* is a variable which is measured on a numeric or quantitative scale, such as the size of population, rate (e.g. rate of population growth), income, age, education (by year) and so on.

However, many social variables are *ordinal variables*, such as attitudes with options such as "very agree", "agree", "not agree nor disagree", "disagree", and "very disagree". Although, an ordinal variable may be turn into a binary variable or a set of dummy variables, it may be considered as a quantitative variable because the statistical efficiency of quantitative variable is more than ordinal variable. The rationale of this is that the results of measuring the relationship between the ordinal variables are very close to the results of measuring it when they are considered quantitative variables.

For example, the Spearman correlation between the two ordinal variables X and Y (with options very low = 1, low = 2, medium = 3, high = 4 and very high = 5) of Table 1.6 is

Table 1.6 Distribution of Y and X

No.	X	Y
1	1	1
2	3	3
3	2	3
4	2	2
5	3	4
6	5	4
7	3	4
8	5	5
9	3	3
10	4	3
11	4	5
12	2	3
13	1	2
Total	38	42

Symmetric Measures

		Value	Asymptotic Standard Error[a]	Approximate T[b]	Approximate Significance
Interval by Interval	Pearson's R	.825	.068	4.833	.001[c]
Ordinal by Ordinal	Spearman Correlation	.828	.090	4.894	.000[c]
N of Valid Cases		13			

a. Not assuming the null hypothesis.

b. Using the asymptotic standard error assuming the null hypothesis.

c. Based on normal approximation.

Fig. 1.17 Symmetric measures of X and Y of Table 1.6. This output is produced by Symmetric Measures Command (see Sect. 1.9)

0.828 (Fig. 1.17). And when they are considered as quantitative variables, the Pearson correlation (Pearson's R) between these two variables is 0.825, which is almost equal to the former. Therefore, many researchers (e.g. Borgatta and Bohrnstedt 1980) acknowledge this rule.

Here, we describe how to examine the other assumptions by using the following example.

Example 1.8 Suppose the data in Table 1.7 are from a probability sample of employees in a town. The independent variables are gender (X1), education (X2) and the dependent variable is (hourly) wage (Y). ◀

Table 1.7 (continued)

No.	X1	X2	Y
28	1	4	49
29	0	3	44
30	1	4	50
31	0	4	40
32	1	4	51
33	0	4	45
34	1	4	55
35	0	4	55
36	1	5	55
37	0	4	60
38	1	5	59
39	0	5	50
40	1	5	61
41	0	5	56
42	1	5	65

(continued)

Table 1.7 Distribution of variables X1, X2 and Y, Example 1.8

No.	X1	X2	Y
1	0	1	10
2	1	1	30
3	0	1	15
4	1	2	25
5	0	1	19
6	1	2	26
7	0	1	20
8	1	2	30
9	0	1	20
10	1	2	35
11	0	1	21
12	1	2	40
13	0	1	24
14	1	3	30
15	0	1	25
16	1	3	36
17	0	2	29
18	1	3	39
19	0	2	30
20	1	3	40
21	0	2	31
22	1	3	45
23	0	2	35
24	1	3	50
25	0	3	36
26	1	4	46
27	0	3	41

(continued)

Table 1.7 (continued)

No.	X1	X2	Y
43	0	5	60
44	1	5	70
45	1	1	30
46	1	5	75
47	0	5	60
48	1	5	75
49	0	5	56
50	1	5	70
51	0	5	50
52	1	5	65
53	0	4	60
54	1	5	61
55	0	4	55
56	1	5	59
57	0	4	45
58	1	5	55
59	0	4	40
60	1	4	55
61	0	3	44
62	1	4	51
63	0	3	41
64	1	4	50
65	0	3	36
66	1	4	49
67	0	2	35
68	1	4	46
69	0	2	31
70	1	3	50
71	0	2	30
72	1	3	45
73	0	2	29
74	1	3	40
75	0	1	25
76	1	3	39
77	0	1	24
78	1	3	36
79	0	1	21
80	1	3	30
81	0	1	20
82	1	2	40

(continued)

Table 1.7 (continued)

No.	X1	X2	Y
83	0	1	20
84	1	2	35
85	0	1	19
86	1	2	30
87	0	1	15
88	1	2	26
89	0	1	10
90	1	2	25
Total		270	3656

Linearity

The most important assumption of linear regression, as its name indicates, is that the dependent variable is a straight-line function of each quantitative independent variable. Moreover, the effects of different independent variables on the dependent variable are additive.

The simple examination of the assumption of linearity is made by the scatterplot of the dependent variable and the independent variable. If the points are almost evenly scattered around a straight line, there is a linear relationship between the dependent variable and the independent variable, such as Fig. 1.18.

Another way to check the linearity is the scatterplot of the standardized residual and the standardized predicted value of the dependent variable. Whenever the relationships between the independent variables and the dependent variable are linear, there isn't any association between the residuals and the predicted values of the dependent variable. In other words, the points are randomly scattered around a horizontal line that passes through the zero point of the vertical axis. In this case, it also indicates that the effects of the independent variables are additive.

For example, the scatterplot of the Standardized Residual and the Standardized Predicted Value in the regression of Y on X1 and X2 in Example 1.8 indicates that there is no pattern or relationship between the residuals and the predicted values (Fig. 1.19).

Fig. 1.18 Scatterplot of wage (Y) and education (X2), Example 1.8. For producing this scatterplot, see Scatterplot Command, Sect. 1.9

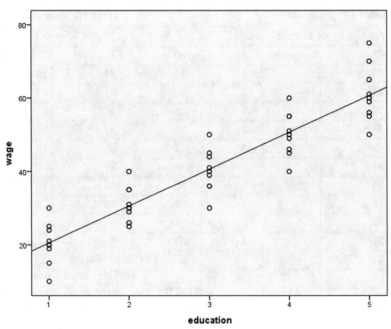

Fig. 1.19 Scatterplot of the Regression Standardized Residual and the Regression Standardized Predicted Value in the regression of Y on X1 and X2, Example 1.8. For producing this scatterplot see, Plot Command Sect. 1.9

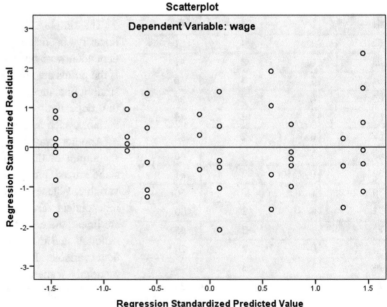

Independence of Residuals

According to the *independence of residuals* assumption, the residuals should not be correlated with each other; they must be independent of one another. To check this assumption, *Durbin-Watson Test* is used. This test values range 0 to 4. If the value of test is around 2, it indicates that the adjacent residuals are not correlated with each other. But if the value of test closes to this or that end values, that is 0 or 4, it shows that adjacent residuals are correlated with each other.

For example, in the regression of the dependent variable Y on the two independent variables X1 and X2 in Example 1.8, Durbin-Watson Test is 1.970 which indicates the lack of correlation between the residuals (see Fig. 1.20, last column).

Normality

Under the *normality* assumption, the distribution of the dependent variable for each value of the independent variable is normal. As a result, the distribution of the residuals must be approximately normal. To check for this assumption, the histogram of the distribution of the standardized residuals is used. If the shape of the histogram remains close to the normal distribution and the residuals are not very skewed the assumption of normality is met.

For example, Fig. 1.21 shows the histogram of the distribution of the standardized residuals of the linear regression of the dependent variable Y on the two independent variables X1 and X2 in Example 1.8. The form of this distribution is close to the normal distribution. Therefore, assuming normal distribution of residuals isn't questioned.

Model Summary[b]

Model	R	R Square	Adjusted R Square	Std. Error of the Estimate	Durbin-Watson
1	.930[a]	.865	.862	5.736	1.970

a. Predictors: (Constant), X2 education, X1 sex

b. Dependent Variable: Y wage

Fig. 1.20 Model summary with Durbin-Watson Test, Example 1.8. This output is produced by Durbin-Watson Command, Sect. 1.9

Fig. 1.21 Histogram of the standardized residuals in the regression of Y on X1 and X2, Example 1.8. For this scatterplot, see Histogram Command, Sect. 1.9

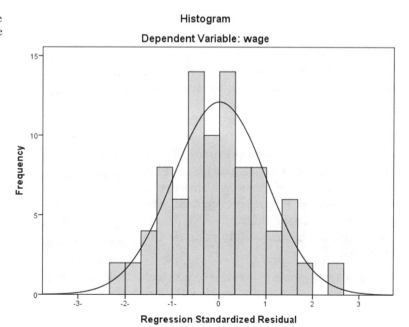

Equality of Variance

According to the *equality of variance* assumption, which is also known as *homogeneity of variance* or *homoscedasticity*, the variance of the distribution of the dependent variable is the same for any combination of values of independent variables. If the spread of the residuals increase or decrease with the predicted values, one must question the assumption of equal variances. Indeed the equality of variance is equivalent to the even distribution of the residuals for each combination of values of the independent variables. So, just like the linearity of the residuals, the equality of variance is examined through the scatterplot of the regression standardized residual with the regression standardized predicted value.

For example, as mentioned previously, the scatterplot of the regression standardized residual and the regression standardized predicted value in the regression of Y on X1 and X2 in Example 1.8 indicates that the spread of hourly wage doesn't increase or decrease with the magnitude of the predicted value (Fig. 1.19); the residuals is randomly scattered around a horizontal line. In other words, there isn't any relationship between residuals and predicted values. Thus, the assumption of equal variances isn't questioned.

1.8.2   Multicollinearity

In social phenomena, the correlation between the independent variables is commonplace. However, when such a correlation is high, which is called *multicollinearity* it may be a problem in the linear regression analysis. For example, when there is a complete correlation between two independent variables in linear regression model, one of them is automatically removed from analysis in SPSS. Or, where there is a high correlation between two or more independent variables, the regression coefficients become unreliable.

This problem is presented by using the following example.

Example 1.9 Suppose Y is affected by X1 and X2 and the data of Table 1.8 are from a probability sample. Also, X3 is a repetition of X2, i.e.

Table 1.8 Distribution of the variables in Example 1.9

No.	X1	X2	X3	X4	Y
1	0	0	0	0	28
2	1	0	0	1	25
3	0	4	4	2	24
4	1	3	3	2	29
5	0	7	7	6	28
6	1	8	8	10	33
8	0	6	6	8	28
8	1	8	8	10	33
9	0	12	12	8	32
10	1	12	12	14	38
11	0	16	16	18	36
12	1	16	16	22	41
Total		92	92	101	375

there is a perfect correlation between these two variables. Therefore, in output of the linear regression of Y on X1, X2 and X3, only X1 and X3 remains in model and the other independent variable (X2) is removed (see Fig. 1.22).

Now, suppose there is a high correlation between X4 and X2 (Fig. 1.23). Therefore, in the output of the linear regression of Y on X1, X2 and X4, the regression coefficients alter and none of them is significant, and as result, the model become unfitted (see Fig. 1.24).

The multicollinearity is problematic because the highly correlated independent variables contain similar information and this tautology makes their effects to be impaired and distorted. The multicollinearity is measured by tolerance. ◀

Tolerance

The linear relationship between an independent variable and the other independent variables is measured by a statistic called tolerance. The *tolerance* of an independent variable is the proportion of its variation that isn't explained by the other independent variables in a given regression model.

The tolerance varies 0 to 1. The tolerance of 0 indicates the perfect correlation between the independent variable in question and the other independent variables. The tolerance of 1

Coefficients[a]

Model		Unstandardized Coefficients		Standardized Coefficients		
		B	Std. Error	Beta	t	Sig.
1	(Constant)	23.281	1.187		19.606	.000
	X1	3.564	1.183	.357	3.012	.015
	X3	.807	.112	.852	7.180	.000

a. Dependent Variable: Y

Fig. 1.22 Coefficients of the regression of Y on X1, X2 and X3, Example 1.9

Correlations

		X2	X4
X2	Pearson Correlation	1	.940[**]
	Sig. (2-tailed)		.000
	N	12	12
X4	Pearson Correlation	.940[**]	1
	Sig. (2-tailed)	.000	
	N	12	12

**. Correlation is significant at the 0.01 level (2-tailed).

Fig. 1.23 Correlation of X2 and X4, Example 1.9

Coefficients[a]

Model		Unstandardized Coefficients		Standardized Coefficients		
		B	Std. Error	Beta	t	Sig.
1	(Constant)	23.989	1.270		18.896	.000
	X1	2.615	1.357	.262	1.927	.090
	X2	.350	.370	.369	.946	.372
	X4	.389	.301	.517	1.293	.232

a. Dependent Variable: Y

Fig. 1.24 Coefficients of the regression of Y on X1, X2 and X4, Example 1.9

Coefficients[a]

Model		Unstandardized Coefficients		Standardized Coefficients			Collinearity Statistics	
		B	Std. Error	Beta	t	Sig.	Tolerance	VIF
1	(Constant)	23.989	1.270		18.896	.000		
	X1	2.615	1.357	.262	1.927	.090	.707	1.415
	X2	.350	.370	.369	.946	.372	.086	11.651
	X4	.389	.301	.517	1.293	.232	.082	12.196

a. Dependent Variable: Y

Fig. 1.25 Coefficients of the regression of Y on X1, X2 and X4 with Collinearity Statistics, Example 1.9. This output is produced by Collinearity Command, Sect. 1.9

indicates the lack of correlation between the independent variable in question and the other independent variables. According to a rule of thumb, if the tolerance is less than 0.10, multicollinearity may be a problem.

For example in the regression of Y on X1, X2 and X4 in Example 1.9, the tolerance of X4 is 0.082 (Fig. 1.25) which indicates that there is a high correlation between X4 and two independent variables X1 and X2.

To avoid the distortion of regression coefficients of highly correlated independent variables, one can remove the variable which appears less important from model, e.g. X4 from above model.

> Multicollinearity refers to a high correlation between two or more independent variables in a multiple linear regression model.

> When the tolerance, as a measure of multicollinearity is close to zero, it indicates a high correlation between an independent variable with the other independent variables.

1.8.3 Unusual Cases

The *unusual case* is an element which is outlier and different from the other elements or is influential case and has a large impact on the regression model.

Here, the outlier and influential cases are discussed by using the following example.

Example 1.10 Assume that the data in this example is the same as in Example 1.8 (Table 1.7), except that two cases 91 and 92 have been added; Table 1.9 presents the distribution of the independent variables X1 (gender), X2 (education) and the dependent variable Y (hourly wage), which are from a probability sample of employees in a town with some of statistics for *diagnosis* of the unusual cases. The linear regression model of this example is the regression of Y on X1 and X2. ◄

Outlier
When the standardized residual value of a case is greater than the absolute value of 3, it is considered *outlier*. For example, as shown in Fig. 1.26, the case 91 and case 92 are obviously outlier; their standardized residuals, 4.085 and −5.612 respectively are absolutely very greater than other standardized residuals which are generally between −2 and 2 (see Table 1.9, column 4).

Influential case
The *influential case* is an element which has a large impact on the results of regression analysis. There are several statistics for diagnosis of the influential cases such as *Standardized Different Fit* (DFFIT) which is standardized difference between the predicted values of the dependent

Table 1.9 Distribution of variables, with statistics for diagnosis of the unusual cases, Example 1.10

Case	X1	X2	Y	ZRE_1	SDF_1	MAH_1	COO_1	LEV_1
Number	(1)	(2)	(3)	(4)	(5)	(6)	(7)	(8)
1	0	1	10	−1.460	−0.283	2.157	0.026	0.024
2	1	1	30	0.470	0.117	4.101	0.005	0.045
3	0	1	15	−0.914	−0.176	2.157	0.010	0.024
4	1	2	25	−0.945	−0.180	2.102	0.011	0.023
5	0	1	19	−0.477	−0.091	2.157	0.003	0.024
6	1	2	26	−0.836	−0.159	2.102	0.008	0.023
7	0	1	20	−0.367	−0.070	2.157	0.002	0.024
8	1	2	30	−0.399	−0.076	2.102	0.002	0.023
9	0	1	20	−0.367	−0.070	2.157	0.002	0.024
10	1	2	35	0.148	0.028	2.102	0.000	0.023
11	0	1	21	−0.258	−0.049	2.157	0.001	0.024
12	1	2	40	0.694	0.132	2.102	0.006	0.023
13	0	1	24	0.070	0.013	2.157	0.000	0.024
14	1	3	30	−1.268	−0.197	1.086	0.013	0.012
15	0	1	25	0.179	0.034	2.157	0.000	0.024
16	1	3	36	−0.612	−0.094	1.086	0.003	0.012
17	0	2	29	−0.253	−0.040	1.162	0.001	0.013
18	1	3	39	−0.284	−0.044	1.086	0.001	0.012
19	0	2	30	−0.144	−0.022	1.162	0.000	0.013
20	1	3	40	−0.175	−0.027	1.086	0.000	0.012
21	0	2	31	−0.034	−0.005	1.162	0.000	0.013
22	1	3	45	0.372	0.057	1.086	0.001	0.012
23	0	2	35	0.403	0.063	1.162	0.001	0.013
24	1	3	50	0.918	0.142	1.086	0.007	0.012
25	0	3	36	−0.357	−0.056	1.151	0.001	0.013
26	1	4	46	−0.388	−0.059	1.055	0.001	0.012
27	0	3	41	0.190	0.030	1.151	0.000	0.013
28	1	4	49	−0.060	−0.009	1.055	0.000	0.012
29	0	3	44	0.517	0.081	1.151	0.002	0.013
30	1	4	50	0.049	0.007	1.055	0.000	0.012
31	0	4	40	−0.789	−0.151	2.124	0.008	0.023
32	1	4	51	0.158	0.024	1.055	0.000	0.012
33	0	4	45	−0.242	−0.046	2.124	0.001	0.023
34	1	4	55	0.595	0.091	1.055	0.003	0.012
35	0	4	55	0.851	0.163	2.124	0.009	0.023
36	1	5	55	−0.274	−0.051	2.008	0.001	0.022
37	0	4	60	1.397	0.269	2.124	0.024	0.023
38	1	5	59	0.164	0.031	2.008	0.000	0.022
39	0	5	50	−0.565	−0.141	4.081	0.007	0.045

(continued)

Table 1.9 (continued)

Case	X1	X2	Y	ZRE_1	SDF_1	MAH_1	COO_1	LEV_1
Number	(1)	(2)	(3)	(4)	(5)	(6)	(7)	(8)
40	1	5	61	0.382	0.071	2.008	0.002	0.022
41	0	5	56	0.091	0.023	4.081	0.000	0.045
42	1	5	65	0.819	0.153	2.008	0.008	0.022
43	0	5	60	0.528	0.131	4.081	0.006	0.045
44	1	5	70	1.366	0.258	2.008	0.022	0.022
45	1	1	30	0.470	0.117	4.101	0.005	0.045
46	1	5	75	1.912	0.365	2.008	0.043	0.022
47	0	5	60	0.528	0.131	4.081	0.006	0.045
48	1	5	75	1.912	0.365	2.008	0.043	0.022
49	0	5	56	0.091	0.023	4.081	0.000	0.045
50	1	5	70	1.366	0.258	2.008	0.022	0.022
51	0	5	50	−0.565	−0.141	4.081	0.007	0.045
52	1	5	65	0.819	0.153	2.008	0.008	0.022
53	0	4	60	1.397	0.269	2.124	0.024	0.023
54	1	5	61	0.382	0.071	2.008	0.002	0.022
55	0	4	55	0.851	0.163	2.124	0.009	0.023
56	1	5	59	0.164	0.031	2.008	0.000	0.022
57	0	4	45	−0.242	−0.046	2.124	0.001	0.023
58	1	5	55	−0.274	−0.051	2.008	0.001	0.022
59	0	4	40	−0.789	−0.151	2.124	0.008	0.023
60	1	4	55	0.595	0.091	1.055	0.003	0.012
61	0	3	44	0.517	0.081	1.151	0.002	0.013
62	1	4	51	0.158	0.024	1.055	0.000	0.012
63	0	3	41	0.190	0.030	1.151	0.000	0.013
64	1	4	50	0.049	0.007	1.055	0.000	0.012
65	0	3	36	−0.357	−0.056	1.151	0.001	0.013
66	1	4	49	−0.060	−0.009	1.055	0.000	0.012
67	0	2	35	0.403	0.063	1.162	0.001	0.013
68	1	4	46	−0.388	−0.059	1.055	0.001	0.012
69	0	2	31	−0.034	−0.005	1.162	0.000	0.013
70	1	3	50	0.918	0.142	1.086	0.007	0.012
71	0	2	30	−0.144	−0.022	1.162	0.000	0.013
72	1	3	45	0.372	0.057	1.086	0.001	0.012
73	0	2	29	−0.253	−0.040	1.162	0.001	0.013
74	1	3	40	−0.175	−0.027	1.086	0.000	0.012
75	0	1	25	0.179	0.034	2.157	0.000	0.024
76	1	3	39	−0.284	−0.044	1.086	0.001	0.012
77	0	1	24	0.070	0.013	2.157	0.000	0.024
78	1	3	36	−0.612	−0.094	1.086	0.003	0.012

(continued)

Table 1.9 (continued)

Case Number	X1 (1)	X2 (2)	Y (3)	ZRE_1 (4)	SDF_1 (5)	MAH_1 (6)	COO_1 (7)	LEV_1 (8)
79	0	1	21	−0.258	−0.049	2.157	0.001	0.024
80	1	3	30	−1.268	−0.197	1.086	0.013	0.012
81	0	1	20	−0.367	−0.070	2.157	0.002	0.024
82	1	2	40	0.694	0.132	2.102	0.006	0.023
83	0	1	20	−0.367	−0.070	2.157	0.002	0.024
84	1	2	35	0.148	0.028	2.102	0.000	0.023
85	0	1	19	−0.477	−0.091	2.157	0.003	0.024
86	1	2	30	−0.399	−0.076	2.102	0.002	0.023
87	0	1	15	−0.914	−0.176	2.157	0.010	0.024
88	1	2	26	−0.836	−0.159	2.102	0.008	0.023
89	0	1	10	−1.460	−0.283	2.157	0.026	0.024
90	1	2	25	−0.945	−0.180	2.102	0.011	0.023
91	0	1	60	4.004	0.850	2.157	0.198	0.024
92	1	8	30	−5.612	−2.990	10.770	1.789	0.118
Mean		3	40.7	0.000	−0.019	1.978	0.027	0.022

ZRE_1 standardized residuals; *SDF_1* Standardized DFFIT; *MAH_1* Mahalanobis Distance; *COO_1* Cook's Distance; *LEV_1* Centered Leverage Value

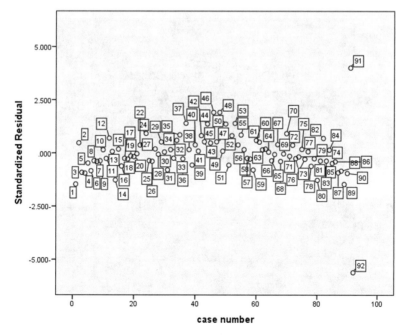

Fig. 1.26 Scatterplot of the standardized residuals and the case number, Example 1.10 (for producing this scatterplot, see Sect. 1.9)

variable when a certain case is included in computation and when it is not.

For example, as shown in Fig. 1.27, the Standardized Different Fit (Standardized DFFIT) of the case 92 (about −2.990) is far from the Standardized Different Fits of the other cases which are generally around zero (see Table 1.9, column 5).

The distance of cases from average values of the independent variables which may influence the regression analysis can be measured by complex statistics including Centered Leverage Values, Mahalanobis Distance, and Cook's Distance.

For example, as shown in Fig. 1.28, the Centered Leverage Value (LEV_1) of the case 92 (about 0.118) is much larger than the *Centered Leverage Values* of other cases which are generally around the mean of the *Centered Leverage Values*, K/n = 2/92 = 0.022 (K is the number of the independent variables and n is the *sample size*) (see Table 1.9, column 8).

Or, the Mahalanobis Distance of the case 92 (about 10.770) is much larger than the *Mahalanobis Distances* of other cases which are generally around the mean of the Mahalanobis Distance,   K(1 − 1/n) = 2(1 − 1/92) = 1.987 (see Table 1.9, column 6).

Also, the Cook's Distance of the case 92 (about 1.789) which is a measure of the change of the residuals of all cases when that case remove from the regression analysis is much larger than the Cook's Distances of other cases which are generally close to zero (see Table 1.9, column 7).

> An unusual case is different from the others or has a large impact on the regression model.

Solution of Unusual Cases

If there are *unusual cases* that are not result of the errors in data and if the remove them leads to increase in the coefficient of determination, it is better to present two models: a model based on

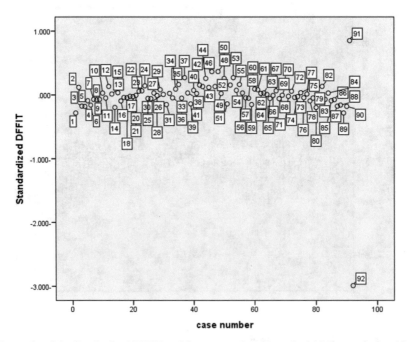

Fig. 1.27 Scatterplot of the Standardized DFFIT and the case number, Example 1.10 (for producing this scatterplots, see Sect. 1.9)

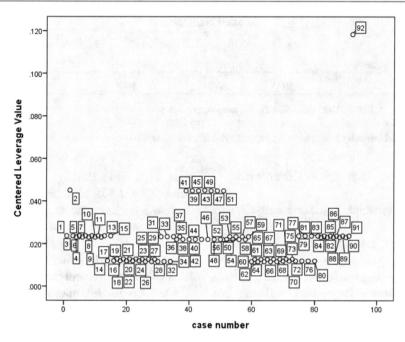

Fig. 1.28 Scatterplot of the Centered Leverage Values and the case number, Example 1.10 (for producing this scatterplots , see Sect. 1.9)

Model Summary

Model	R	R Square	Adjusted R Square	Std. Error of the Estimate
1	.810[a]	.656	.648	9.151

a. Predictors: (Constant), X2 education, X1 sex

Fig. 1.29 Model summary, Example 1.10

all cases and another without unusual cases. The former model is necessary for professional commitment to present perfectly facts and collected data. The latter model is to present an accurate picture of the common pattern of reality. Anyway, if possible, one must try to explain the cause or causes of unusual cases.

For example, after removing the cases 91 and 92, the coefficient of determination increases by 0.209 (from 0.656 to 0.865; see Figs. 1.29 and 1.30).

The remove unusual cases can be limited to a maximum of 5% of all cases. Thus, the regression model includes at least 95% of the cases.

1.9   Commands

Here, the SPSS[12] commands for the calculations of multiple regression analysis in this chapter are presented: Linear Regression Command, Correlation Command, Dummy Variable Command, Compare Mean Command, Symmetric Measures Command, Histogram Command, Durbin-Watson Command, Collinearity Command, Save Command, Scatterplot Command, and Plot Command.

---

[12]Statistics Package for Social Sciences.

Model Summary

Model	R	R Square	Adjusted R Square	Std. Error of the Estimate
1	.930[a]	.865	.862	5.736

a. Predictors: (Constant), X2 education, X1 sex

Fig. 1.30 Model summary after removing cases 91 and 92, Example 1.10

1.9.1 Linear Regression Command

To carry out a multiple regression analysis, after opening the data file:

1. In the Data Editor as shown in Fig. 1.31, select
   Analyze → Regression → Linear…,
2. In the **Linear Regression** dialogue box (Fig. 1.32), select the dependent variable (e.g. Y) from the list of variables and transfer it into the **Dependent:** box and the independent variables (e.g. X1, X2 and X3) into the **Independent(s):** box,
3. Click on the **OK** button.

1.9.2 Correlation Command

To obtain a correlation between two variables Y and X:

1. In the Data Editor (data file) as shown in Fig. 1.33, select
   Analyze → Correlate → Bivariate…,
2. In the **Bivariate Correlations** dialogue box (Fig. 1.34), select the dependent variable (e.g. Y) and the independent variable (e.g. X2) from the list of variables and transfer it into the **Variables:** box,
3. Click on the **OK** button.

1.9.3 Dummy Variable Command

To create a set of dummy variables from a qualitative variable with more than two categories:

1. In the Data Editor (data file) as shown in Fig. 1.35, select **Transform → Create Dummy Variables**,

2. In the **Create Dummy Variables** dialogue box (Fig. 1.36), from the list of **variables**, choose the independent variable from which you want create the dummy variables (e.g. Race [X1]), and transfer it into **Create Dummy Variables for:** box,
3. In the **Measurement Level Usage** section, select **Create dummies for all variables**,
4. In the **Root Names (One Per Selected Variable)** section, specify names to be used as the prefix of all dummy variables for the selected variable (e.g. RACE),
5. Click **OK**.

1.9.4 Compare Mean Command

To obtain the mean of a quantitative dependent variable by various categories of a qualitative independent variable:

1. In the Data Editor as shown in Fig. 1.37, select
   Analyze → Compare Mean → Means …,
2. In the **Means** dialogue box (Fig. 1.38), select the dependent variable from the list of variables and transfer it into the **Dependent List:** box and the independent variable into the **Independent List:** box,
3. Click on the **OK** button.

1.9.5 Symmetric Measures Command

To obtain a Symmetric Measures table for the correlation between two variables:

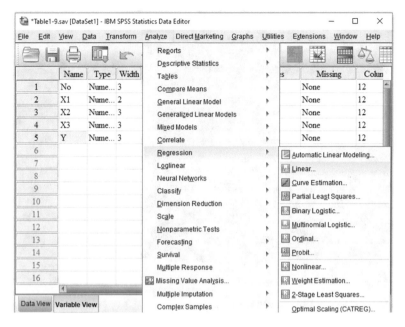

Fig. 1.31 Data Editor, Analyze, Regression, Linear, …

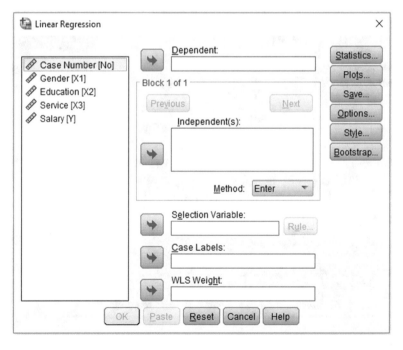

Fig. 1.32 Linear Regression dialog box

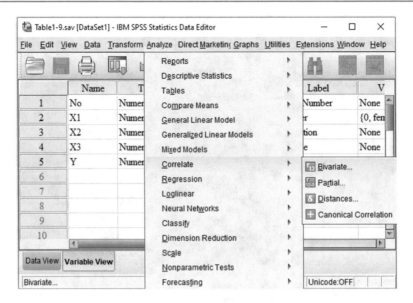

Fig. 1.33 Data Editor, Analyze, Correlate, Bivariate, …

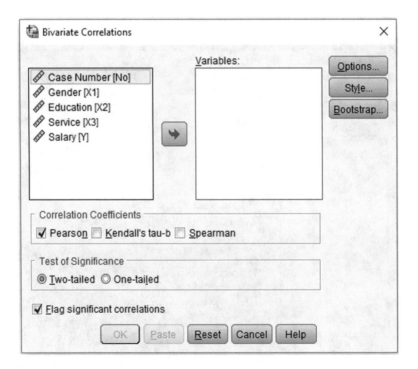

Fig. 1.34 Bivariate Correlation dialogue box

Fig. 1.35 Data Editor, Transform, Create Dummy Variables

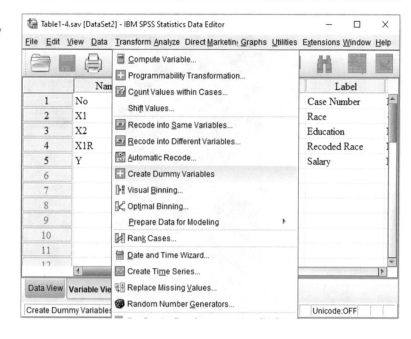

Fig. 1.36 Create Dummy Variables dialogue box

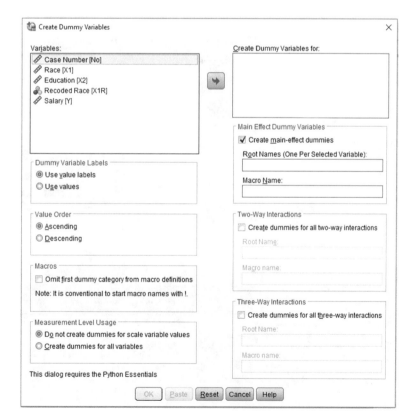

Fig. 1.37 Data Editor,
Analyze, Compare Mean,
Means, …

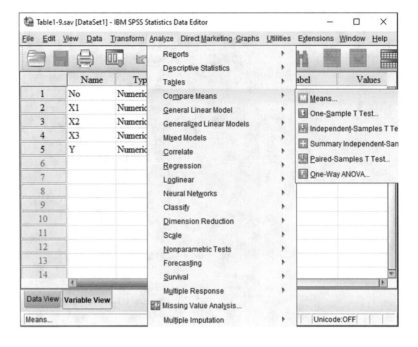

Fig. 1.38 Means dialog box

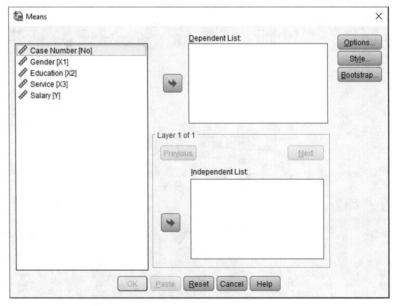

1. In the Data Editor (data file) as shown in Fig. 1.39, select
   Analyze → Descriptive Statistics → Crosstabs…,
2. In the **Crosstabs** dialogue box (Fig. 1.40), select one variable from the list of variables and transfer it into the **Row(s)**: box and other variable into the **Column(s)**: box, then click on **Statistics…**,
3. In the **Crosstabs: Statistics** dialogue box (Fig. 1.41), click on **Correlations**, then on the **Continue** button,
4. Click **OK**.

1.9.6 Histogram Command

To obtain the Histogram of the standardized residuals:

1. In the **Linear Regression** dialogue box (Fig. 1.32) after selecting variables, click on the **Plots…** button,
2. In the **Linear Regression: Plots** dialogue box (Fig. 1.42), click on the **Histogram** section,
3. Click on the **Continue** button, then on **OK**.

1.9.7 Durbin-Watson Command

To obtain Durbin-Watson Test:

1. In the **Linear Regression** dialogue box (Fig. 1.32) after selecting variables, click on the **Statistics …** button,
2. In the **Linear Regression: Statistics** dialogue box (Fig. 1.43), click on the **Durbin-Watson**,
3. Click on the **Continue** button,
4. In the **Linear Regression** dialogue box, click **OK**.

1.9.8 Collinearity Command

To obtain the collinearity statistics:

1. In the **Linear Regression** dialogue box (Fig. 1.32) after selecting variables, click on the **Statistics …** button,

2. In the **Linear Regression: Statistics** dialogue box (Fig. 1.43), click on the **Collinearity Diagnostics**,
3. Click on the **Continue** button, then the **OK**.

1.9.9 Save Command

To create the Standardized Residual, and the diagnostics statistics, such as DFBETA, Standardized DFFIT, Centered Leverage Values, Mahalanobis Distance, and Cook's Distance in order to obtain Scatterplot of each of them as variables and another variable such as case number:

1. In the **Linear Regression** dialogue box (Fig. 1.32) after selecting variables, click on the **Save…** button,
2. In the **Linear Regression: Save** dialogue box (Fig. 1.44), select statistics (for example, in **Residuals** section, click on **Standardized**; or in **Influence Statistics** section, click on **Standardized DiFit**; or in **Distances** section, click on **Leverage Values**),
3. Click on the **Continue** button, then on **OK**.

Then, to obtain the scatterplot of each of these statistics as a variable and the case number (a variable includes the numbers of cases), use Scatterplot Command, Sect. 1.9.10.

1.9.10 Scatterplot Command

To obtain the scatterplot of two variables such as the dependent variable and an independent variable:

1. In the Data Editor (data file) as shown in Fig. 1.45, select
   Graphs → Legacy Dialog → Scatter/Dot…,
2. In the **Scatter/Dot** dialogue box (Fig. 1.46), click on **Simple Scatter** and then on the **Define** button,
3. In the **Simple Scatterplot** dialogue box (Fig. 1.47), from the list of variables, select the dependent variable and transfer it into **Y Axis:** box and the independent variable into **X Axis**: box,

Fig. 1.39 Data Editor,
Analyze, Descriptive
Statistics, Crosstabs, …

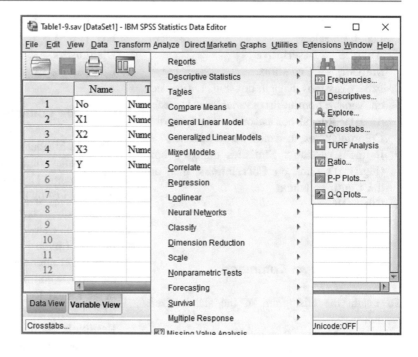

Fig. 1.40 Crosstabs dialog
box

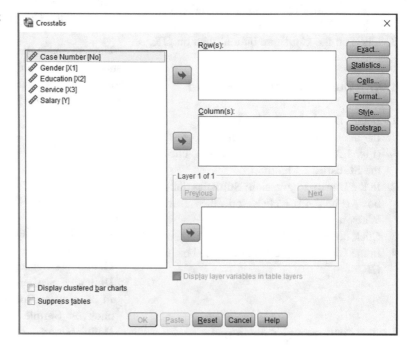

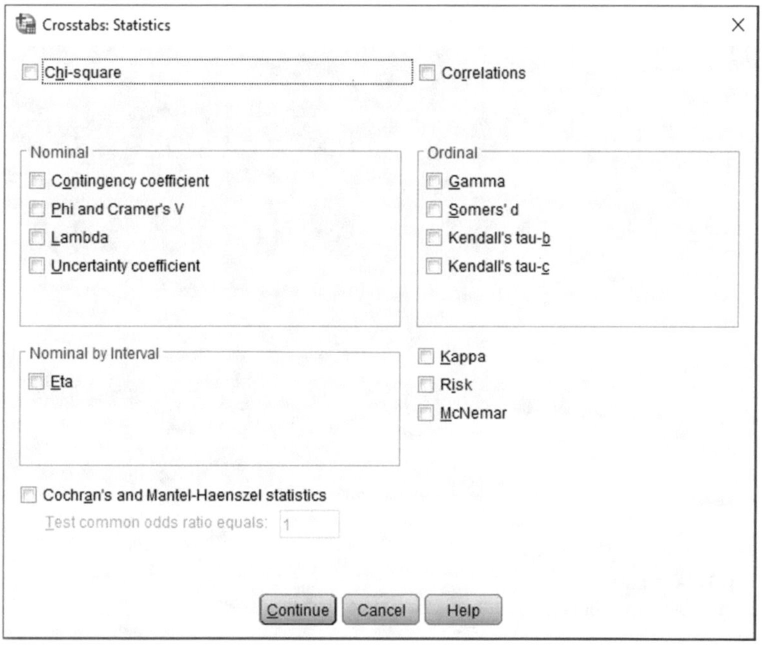

Fig. 1.41 Crosstabs: statistics dialog box

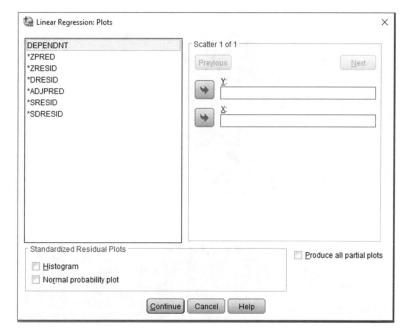

Fig. 1.42 Linear Regression: plots dialogue box

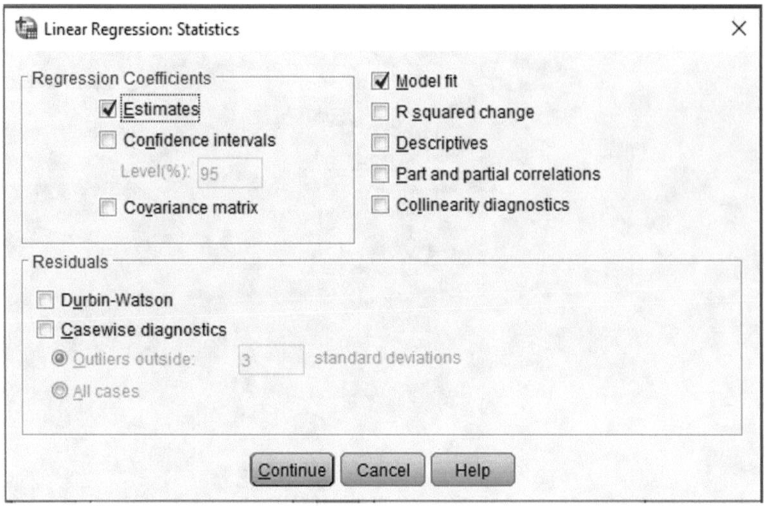

Fig. 1.43 Linear
Regression: statistics dialogue
box

Fig. 1.44 Linear
Regression: save dialogue
box

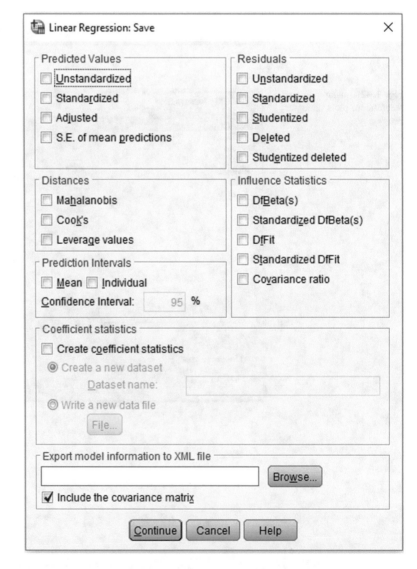

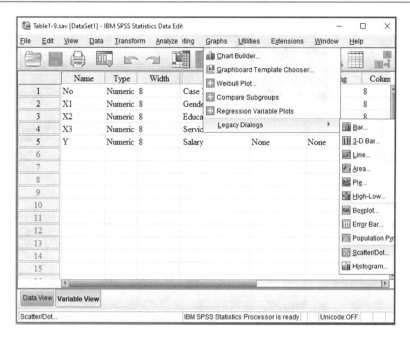

Fig. 1.45 Data Editor, Graphs, Legacy Dialog, Scatter/Dot, …

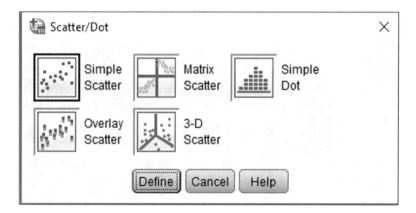

Fig. 1.46 Scatter/dot dialogue box

4. Click on the **OK** bottom,
5. In the **Output** (Fig. 1.48) double click on the Scatterplot in order to activate it,
6. In the **Chart Editor** (Fig. 1.49) click on 📈 (Add Fit Line at Total) in order to add fit line,

7. In the **Properties** dialogue box (Fig. 1.50) delete tick on the **Attach label to line** section and then click on **Apply** button in order to delete the equation on the fit linear,
8. Close **Chart Editor**.

Fig. 1.47 Simple scatterplot
dialogue box

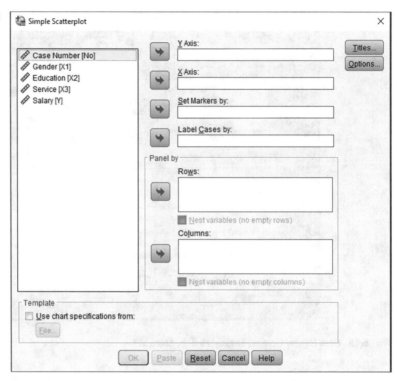

Fig. 1.48 Output of
scatterplot

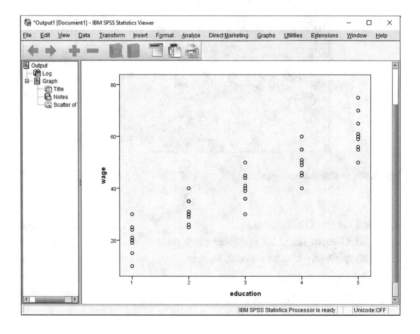

Fig. 1.49 Chart Editor

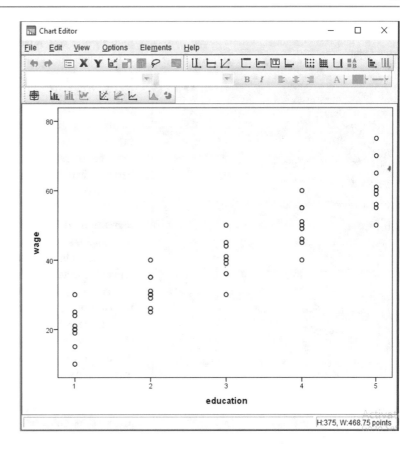

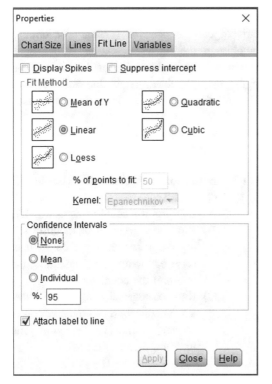

Fig. 1.50 Properties dialogue box

Case Label Command To specify the number of each case (point) in the diagram such as Fig. 1.26:

1. In the **Chart Editor** (Fig. 1.49) right-clicks twice on one of the points,
2. In the Case label Command dialogue box (Fig. 1.51) click on the **Show Data Labels** section.

1.9.11   Plots Command

To obtain the scatterplot of the Regression Standardized Residual and the Regression Standardized Predicted Value:

1. In the **Linear Regression** dialogue box (Fig. 1.32) after selecting variables, click on the **Plots…** button,
2. In the **Linear Regression: Plots** dialogue box (Fig. 1.42), select the ***ZRESID** from the list of variables and transfer it into the **Y:** box and the ***ZPRED** into the **X:** box,

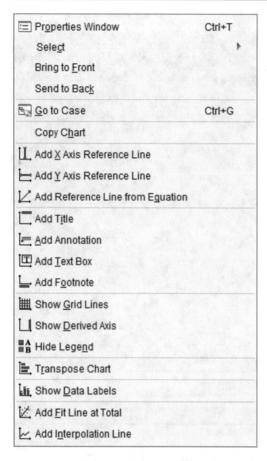

Fig. 1.51 Case label command dialogue box

3. Click on the **Continue** button,
4. In the **Linear Regression** dialogue box (Fig. 1.32), click **OK**,
5. In the **Output** (Fig. 1.48) double click on the Scatterplot in order to activate it,
6. In the **Chart Editor** (Fig. 1.49) click on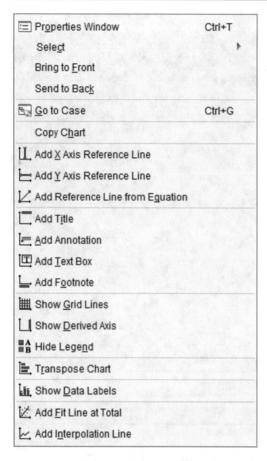

   (Add Fit Line at Total) in order to add fit line,
7. In the **Properties** dialogue box (Fig. 1.50) delete tick on the **Attach label to line** section and then click on **Apply** button in order to delete the equation on the fit linear,
8. Close **Chart Editor**.

1.10 Summary

In this chapter, the multiple linear regression is described by a nontechnical language and simple hypothetical and real examples. It is shown how the multiple linear regression can predict and recreate the real relationships between the variables (phenomena) as a regression equation. This regression equation illustrates how the dependent variable (effect) is related to each of the independent variables (causes) as regression coefficients. It is given a brief overview of how the regression coefficients don't change by the entering or removing of the uncorrelated independent variables, but change by of the uncorrelated ones. It has been presented that the assumed causal relationships are confirmed if all regression coefficients are nonzero, all observed relationships of the independent variables with the dependent variable correspond to the theoretical expectations, and all coefficients are significant when the data are collected from a probability sample. In a fitted linear regression model, the contribution of the independent variables to the variation of the dependent variable is presented by the coefficient of determination which is measure of goodness of fit, too. The regression coefficient of each independent variable is based on controlling for the other independent variables or removing the effects of the other independent variables. It is emphasized that the appropriate regression method for testing hypotheses is the enter method. The qualitative variable can be used as a binary or dummy variable in the linear regression. The chapter concludes with a discussion of the assumptions of the multiple linear regression and how can deal with the multicollinearity and unusual cases.

1.11 Exercises

(1.1) Based on the data of Example 1.4 (Table 1.2):
   (A) Interpret the coefficients in Eq. 1.7.
   (B) Run the regression of monthly salary on gender and the duration of service, write its equation and interpret the coefficients.
   (C) Run the regression of monthly salary on education and the duration of service, write its equation and interpret the coefficients.

(1.2) In which of above models (A, B and C) the coefficients of the independent variables are the same as the salary coefficients in the firm B? Why?

(1.3) In which of above models the coefficients of the independent variables doesn't accurately reflect the salary coefficients in the firm B? Why?

(1.4) In which of above models the coefficients doesn't change? Why?

(1.5) Which of above models is a fitted regression model?

(1.6) Which of above models is a better model? Why?

(1.7) Interpret the coefficient of determination of that model.

(1.8) Write equation of the regression of occupational prestige on sex and education in Example 1.5 (see Fig. 1.10) and interpret its coefficients.

(1.9) Test the hypotheses of Example 1.5 by using the data of the General Social Survey 2018 which is a probability sample of US people. Is the regression of occupational prestige on sex and education a fitted regression model?

(1.10) Refer to the data of Table 1.10, turn race (X1) into a binary variable and run the regression of income (Y) on the above new created independent variable and education (X2).
(A) Is it a fitted regression model?
(B) Write equation of above regression model.
(C) Interpret the regression coefficients.

(1.11) Again, refer to the data of Table 1.10, turn race (X1) into a set of dummy variables. How many dummy variables did you create? Why?

(1.12) Run the regression of independent variable Y on the above dummy variables and X2.
(A) Is it a fitted regression model?
(B) Write equation of above regression model
(C) Interpret the regression coefficients.

(1.13) Which of above two models is a fitted regression model?

(1.14) Test the hypotheses of Example 1.7 using the data of the General Social Survey 1972 which is a probability sample of US people. Turn RACE into a binary variable and run the regression of PRESTIGE on the above new created independent variable and EDUC.
(A) Is it a fitted regression model?
(B) Write equation of above regression model
(C) Interpret the regression coefficients.

(1.15) Now, turn RACE into a set of dummy variables. How many dummy variables did you create? Why?

(1.16) Run the regression of the dependent variable (PRESTIGE) on the above dummy variables and X2. Is it a fitted regression model? Why?

(1.17) Given the data of Table 1.11:
(A) Is there a linear relationship between Y and X1?
(B) Is there a linear relationship between Y and X2?
(C) Is the assumption of the independence of residuals met?
(D) What can you say about the assumption of the normality?
(E) What about the assumption of the equality of variance?
(F) Is the multicollinearity a problem here?
(G) Is there any unusual case?
(H) If there is, what do you do with it?

(1.18) Is the regression of optimism (Y) on education (X1) and job rank (X2) in the data of Table 1.11 a fitted regression model?

(1.19) Write regression equation of above model and interpret the coefficients.

(1.20) What is the meaning of the coefficient of determination in above regression model?

Table 1.10 Distribution of race, education and income of the people in a town

No.	X1 Race[a]	X2 Education	Y Income
1	3	0	1800
2	3	2	2000
3	3	10	2200
4	2	2	2400
5	2	4	3000
6	2	10	3400
7	1	2	3200
8	1	4	3200
9	1	6	4000
10	1	12	4800
11	3	0	1800
12	3	2	2000
13	2	2	2400
14	1	2	3200
15	2	4	3000
16	1	4	3200
17	1	6	4000
18	3	10	2200
19	2	10	3400
20	1	12	4800
Total		104	60,000

[a]1 = race A, 2 = race B, 3 = race C

Table 1.11 Distribution of education (X1), job rank (X2) and optimism (Y) in a probability sample of people in a town

No.	X1	X2	Y
1	0	1	1
2	1	1	2
3	2	1	1
4	2	2	3
5	2	2	4
6	3	5	4
7	3	5	5
8	3	5	6
9	4	1	3
10	4	2	4
11	4	4	5
12	6	3	4
13	6	3	5
14	6	4	6
15	6	6	7

(continued)

Table 1.11 (continued)

No.	X1	X2	Y
16	6	6	8
17	8	3	5
18	8	4	5
19	8	6	9
20	8	6	7
21	8	7	8
22	10	4	9
23	10	7	9
24	10	7	10
25	12	8	11
26	12	8	10
27	12	9	11
28	12	9	12
29	12	10	12
30	12	11	13
31	16	10	14
32	16	12	19
Total	232	172	232
Mean	7.25	5.38	7.25

References

Borgatta EF, Bohrnstedt GW (1980) Levels of measurement, once over again. Sociol Res 9(2):147–160
Kerlinger FN, Pedhazur EJ (1973) Multiple regression in behavioral research. Holt Rinehart, Winston, New York

Further Reading

Books on linear regression are numerous. Here are a few ones written with rather simple language:
Allison PD (1999) Multiple regression, a primer. Sage, Thousand Oaks, CA. It is also a good introduction to regression with emphasis on causality
Lewis-Beck M (1980) Applied regression, an introduction. Sage, Newbury Park, CA. It is a good introduction to the regression and its assumptions
These books are more advanced:
Achen CH (1982) Interpreting and using regression. Sage, Beverly Hills, CA. It introduce the informal norms govern regression analysis with emphasis on applications

Best H, Wolf C (eds) (2015) The SAGE handbook of regression analysis and causal inference. Sage, London. It covers a series of good articles on the regression analysis
Chatterjee S, Simonoff JS (2013) Handbook of regression analysis. Wiley, Hoboken, NJ. It presents the various regression techniques including linear logistic and nonlinear regressions
Cohen J, Cohen P, West SG, Aiken LS (2003) Applied multiple regression/correlation analysis for the behavioral sciences. Erlbaum, Mahwah, NJ. It is a wideranging book with emphasis on graphical presentations
Darlington RB, Hayes AF (2017) Regression analysis and linear models, concepts, applications and implementation. Guilford Press, New York. It emphasizes the conceptual understanding with a full chapter (Chap. 8) about assessing the importance of the independent variables
Draper NR, Smith H (1998) Applied regression analysis, 3rd ed. Wiley, New York. It is an introduction to the fundamentals of regression analysis
Fox J (2016) Applied regression analysis and generalized linear models, 3rd ed. Sage, Thousand Oaks, CA. It is a good wide-ranging book about regression analysis and closely related methods

Montgomery DC, Peck EA, Vining GG (2012) Introduction to linear regression analysis, 5th ed. Wiley, Hoboken. It is a wide-ranging and technical book

Olive DJ (2017) Linear regression. Springer, New York, NY. It is a technical book

Ryan TP (1997) Modern regression methods. Wiley, New York. It is a wide-ranging and technical book

These books are mainly about problems of violations of regression assumptions and options for dealing with them:

Atkinson A, Riani R (2000) Robust diagnostic regression analysis. Springer, New York, NY

Belsly DA, Kuh E, Welsch RE (1980) Regression diagnostics, identifying influential data and sources of collinearity. Wiley, Hoboken

Berry WD, Feldman S (1985) Multiple regression in practice. Sage, Thousand Oaks, CA

Berry WD (1993) Understanding regression assumptions. Sage, Newbury Park, CA

Fox J (1991) Regression diagnostics. Sage, Newbury Park, CA

Path Analysis

This chapter shows how a path analysis can test the causal relationships both between a set of independent variables and the dependent variable, and between the independent variables. It presents that the stages of a path analysis involves examining the technical assumptions, testing path diagram, and illustrating the various effects. It also shows how an independent variable has three actual effects on the dependent variable (raw, direct and pure effects) and may have two potential (indirect and spurious) effects. And it presents what is the unique contribution of each independent variable on the dependent variable.

The *path analysis* is a powerful statistical technique to test a causal model which based on theoretical reasoning explains causal relationships between a set of independent variables and the dependent variable. Also, it tests the assumed relationships between the independent variables. Indeed, path analysis is perfectly based upon theoretical reasoning about causal relationships between variables. In this way, we can show all kinds of effects of each independent variable on the dependent variable. And above all, we can show what independent variable is the most effective, important factor. It is only path analysis that can show the unique effect of each independent variable on the dependent variable.

2.1 Causal Model

A *causal model* is a theoretical model that describes the causal mechanisms of a set of independent variables and the dependent variable. The path analysis is restricted to the *recursive causal models* in which no feedback effects are assumed, i.e. all causal relationships are one way. In the path analysis the hypotheses of causal model are incorporated into the multiple linear regression equations. Then, the regression coefficients are calculated in the usual way. A causal model is expressed in the form of a path diagram.

2.2 Path Diagram

A *path diagram* is diagrammatic expression of a causal model, such as Fig. 2.1. In a path diagram, variables are arranged by the causal order from left to right. All independent (causal) variables are connected by arrows to the dependent variable and each independent variable is connected by an arrow to another each independent variable on which has causal effect. The arrowheads delineate the direction of these causal relationships. Indeed every arrow indicates one hypothesis about causal connection between two variables.

H. Nayebi, *Advanced Statistics for Testing Assumed Causal Relationships*, University of Tehran Science and Humanities Series, https://doi.org/10.1007/978-3-030-54754-7_2

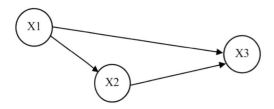

Fig. 2.1 Path diagram of three variables

> A causal model consists of a set of independent variables (causes) and a dependent variable (effect) and their causal relationships.

For example, the path diagram in Fig. 2.1 indicates that two independent variables X1 and X2 have causal effects on the dependent variable X3. In addition, the independent variable X1 has causal effect on the variable X2, too. Thus, this path diagram involves three hypotheses: X1 affects X3; X2 affects X3; and X1 affects X2.

A path analysis involves three stages:
1. Examining technical assumptions,
2. Testing path diagram,
3. Identifying various effects.

In next sections, each of these stages of a path analysis is examined separately.

2.3 Examining Technical Assumptions

Since path analysis is based on multiple linear regression equations, its *technical assumptions* are same ones of multiple linear regression analysis:
1. The independent variables can include quantitative variables and binary variables[1] but the dependent variable is only quantitative variable.
2. The relationship between the dependent variable and each independent variable is linear and additive.

---

[1]The dummy variables should not be used in path analysis, because they are categories of one variable. Hence, they are highly correlated, but their correlations are not causality that can be entered into the causal model.

3. The observations are independent of each other.
4. The distribution of the dependent variable for each value of the independent variable is normal.
5. The variance of the dependent variable is the same for all values of the independent variable.

Also, other considerations of linear regression (multicollinearity and unusual cases) should be considered. Moreover, as mentioned previously path analysis assumes the causal model is a recursive causal model without any feedback effects.

After the collection data from the population (or a probability sample), one must examine the assumptions of path analysis and ensure that there are not severe violation of them (for details of the linear regression considerations, see Sect. 1.8).

Then, it turns to test the hypotheses of the causal model, the test of the path diagram.

2.4 Testing Path Diagram

The *test of a path diagram* is to test its hypotheses. There are two kinds of hypotheses in a path diagram: explicit and implicit. An *explicit hypothesis* states a variable has causal effect on another variable. For example, the path diagram in Fig. 2.1 involves only explicit hypotheses: X1 affects X3; X2 affects X3; and X1 affects X2.

An *implicit hypothesis* indirectly indicates the lack of causal relation between two independent variables. For example, path diagram in Fig. 2.2 doesn't assume any causal relationship between X1 and X2. Thus, it involves two explicit hypotheses (X1 affects X3 and X2 affects X3) and one implicit hypothesis (lack of causal relationship between X1 and X2). To test the implicit hypothesis requires to show that the correlation between two variables is zero or isn't significant.

However, testing the explicit hypotheses requires as much linear regression as the number of the endogenous variables plus one for the dependent variable. In a path diagram, an

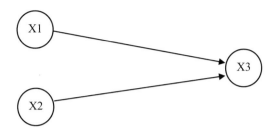

Fig. 2.2 Path diagram of three variables

independent variable that is affected by the other independent variable (or variables) is called *endogenous variable*. For example, in Fig. 2.1 the variable X2 which is affected by the variable X1 is an endogenous variable. An independent variable that isn't affected by any independent variable in the model called an *exogenous variable*, such as X1 in Fig. 2.1 on which no independent variable has effect.

This naming is important in path analysis, since as stated above each endogenous variable requires a linear regression equation to test the hypotheses of path diagram. Moreover, the various effects of these two kinds of independent variables aren't same (see Sect. 2.5).

> An endogenous variable is an independent variable that is affected by the other independent variables in a path diagram.

For example, to test the explicit hypotheses of path diagram of Fig. 2.1 requires two linear regression: (1) the regression of the dependent variable X3 on the independent variables X1 and X2; and (2) the regression of the endogenous variable X2 on the independent variable X1.

Where the coefficients in all linear regressions were nonzero and significant (when data are collected from a probability sample), and all explicit and implicit hypotheses were confirmed (all observed relationships correspond to the theoretical expectations), that path diagram is a fitted one, a *fitted causal model*.

The fitness index of a fitted path diagram, which indicates how well it fits a given facts, is the coefficient of determination in the regression

of the dependent variable on the independent variables.

If a path diagram is fitted, it turns to present the various effects of the independent variables on the dependent variable.

Here, the test of a path diagram is illustrated by using three examples (Examples 2.1–2.3).

> An exogenous variable is an independent variable that isn't affected by any independent variable in a path diagram.

Example 2.1
A researcher assumed that two independent variables of X1 and X2 have causal effect on the dependent variable X3. Also, X1 has a causal effect on X2. This causal model is expressed as a path diagram presented in Fig. 2.1. The appropriate data for test of this path diagram has collected from a probability sample of people in a town (Table 2.1). ◄

Moreover, suppose the examination of the technical assumptions of path analysis indicates that there isn't severe violation of them[2]. Now, it turns to test the hypotheses of path diagram. For this purpose, we must carry out the appropriate linear regression equations. In this example with one endogenous variable, path analysis requires two linear regression equations: (1) the regression of X3 on X1 and X2, (2) the regression of X2 on X1.

The output of the first linear regression shows that the dependent variable has significant relationship with both independent variables (Fig. 2.3). Also, the result of the second linear regression shows that the endogenous variable X2 had significant relationship with exogenous variable X1 (Fig. 2.4).

> Test of a path diagram involves as much linear regressions as the number of endogenous variables plus one.

---

[2]Can you check these assumptions? (see Sect. 1.8).

Table 2.1 Distribution of three variables X1, X2 and X3, Example 2.1

No.	X1	X2	X3
1	0	0	20
2	1	4	29
3	0	0	20
4	1	4	30
5	1	4	32
6	0	0	21
7	1	8	33
8	0	0	23
9	1	8	34
10	1	8	36
11	0	4	24
12	1	12	37
13	1	12	38
14	0	4	27
15	1	12	40
16	1	16	41
17	0	8	29
18	1	16	42
19	0	8	31
20	1	16	44
Total	12	144	631

Thus, all hypotheses of the path diagram Example 2.1 are confirmed. So, we can conclude that the path diagram of Example 2.1 fits data. In other words, the path diagram of Example 2.1 is a fitted one.

The fitness index of this fitted path diagram which is the coefficient of determination of first linear regression equation is 0.970 (Fig. 2.5). It indicates that 97% of variation of X3 is explained by X1 and X2.

Unfitted Path Diagram

There are two cases in which a path diagram isn't confirmed and as a result it isn't a fitted causal model:

1. There is a significant relationship, while path diagram doesn't assume it, i.e. when at least an implicit hypothesis isn't confirmed (such as Example 2.2),
2. At least one of variables of path diagram has not the assumed relationship, i.e. when at least an explicit hypothesis isn't confirmed (such as Example 2.3).

In such cases the causal model isn't verified and considers a wrong model and path analysis isn't continued to next stage.

Example 2.2

A researcher assumed that two independent variables X1 and X2 have causal effects on the dependent variable X3. But she or he doesn't assume any causal relationship between X1 and X2. This causal model is expressed as the path diagram presented in Fig. 2.2. The appropriate data to test this path diagram has collected from a probability sample of people in a town (Table 2.2). In this example with no endogenous variable, the path analysis requires only one

Coefficients[a]

Model		Unstandardized Coefficients		Standardized Coefficients		
		B	Std. Error	Beta	t	Sig.
1	(Constant)	21.274	.537		39.609	.000
	X1	4.724	.823	.317	5.741	.000
	X2	1.034	.076	.750	13.606	.000

a. Dependent Variable: X3

Fig. 2.3 Coefficients of the regression of X3 on X1 and X2, Example 2.1 [This output and two next outputs are produced by Linear Regression Command (see Sect. 1.9).]

Coefficients[a]

Model		Unstandardized Coefficients		Standardized Coefficients	t	Sig.
		B	Std. Error	Beta		
1	(Constant)	3.000	1.509		1.988	.062
	X1	7.000	1.948	.646	3.593	.002

a. Dependent Variable: X2

Fig. 2.4 Coefficients of the regression of X2 on X1, Example 2.1

Fig. 2.5 Model summary, Example 2.1

Model Summary

Model	R	R Square	Adjusted R Square	Std. Error of the Estimate
1	.985[a]	.970	.966	1.376

a. Predictors: (Constant), X2, X1

Table 2.2 Distribution of three variables X1, X2 and X3, Example 2.2

No.	X1	X2	X3
1	0	0	0
2	1	4	9
3	1	4	10
4	1	4	12
5	1	8	13
6	0	0	3
7	1	8	14
8	1	8	16
9	0	4	4
10	1	12	17
11	0	4	7
12	1	16	22
Total	8	72	127

linear regression equation: the regression of X3 on X1 and X2. ◄

> A path diagram is a fitted one where all its explicit and implicit hypotheses were confirmed.

The output of this regression shows that the dependent variable has significant relationship with both independent variables (Fig. 2.6). But, the test of the implicit assumption of no causal relationship between X1 and X2 requires seeing whether there isn't correlation between them. Unlike the above model, Fig. 2.7 shows that these two variables are correlated with each other. In other words, its implicit hypothesis isn't confirmed.

Therefore, this path diagram isn't a fitted one; this causal model isn't verified.

Coefficients[a]

Model		Unstandardized Coefficients		Standardized Coefficients		
		B	Std. Error	Beta	t	Sig.
1	(Constant)	1.611	.774		2.080	.067
	X1	4.958	1.162	.382	4.268	.002
	X2	.944	.122	.691	7.713	.000

a. Dependent Variable: X3

Fig. 2.6 Coefficients of the regression of X3 on X1 and X2, Example 2.2

Fig. 2.7 Correlation of X1 and X2, Example 2.2 [This output is produced by Correlation Command (see Sect. 1.9).]

Correlations

		X1	X2
X1	Pearson Correlation	1	.632[*]
	Sig. (2-tailed)		.027
	N	12	12
X2	Pearson Correlation	.632[*]	1
	Sig. (2-tailed)	.027	
	N	12	12

*. Correlation is significant at the 0.05 level (2-tailed).

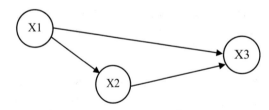

Fig. 2.8 Path diagram of Example 2.3

Example 2.3

A researcher assumed that two independent variables X1 and X2 have causal effects on the dependent variable X3. Also, X1 has a causal effect on X2 (Fig. 2.8). The appropriate data to test this path diagram has collected from a probability sample of people in a town (Table 2.3). ◄

Table 2.3 Distribution of three variables X1 to X3, Example 2.3

No.	X1	X2	X3
1	0	0	20
2	1	0	25
3	0	4	22
4	1	4	29
5	0	8	28
6	1	8	33
7	0	8	28
8	1	8	34
9	0	12	32
10	1	12	37
11	0	16	36
12	1	16	38
Total	6	96	362

Coefficients[a]

Model		Unstandardized Coefficients		Standardized Coefficients		
		B	Std. Error	Beta	t	Sig.
1	(Constant)	20.067	.695		28.860	.000
	X1	5.000	.663	.447	7.542	.000
	X2	.950	.064	.877	14.800	.000

a. Dependent Variable: X3

Fig. 2.9 Coefficients of the regression of X3 on X1 and X2, Example 2.3

Coefficients[a]

Model		Unstandardized Coefficients		Standardized Coefficients		
		B	Std. Error	Beta	t	Sig.
1	(Constant)	8.000	2.309		3.464	.006
	X1	.000	3.266	.000	.000	1.000

a. Dependent Variable: X2

Fig. 2.10 Coefficients of the regression of X2 on X1, Example 2.3

In this example with one endogenous variable, path analysis requires two linear regression equations: the regression of X3 on X1 and X2, and the regression of X2 on X1.

The output of the regression of X3 on X1 and X2 shows that the dependent variable has significant relationship with both independent variables (Fig. 2.9). But, the result of the regression of X2 on X1 shows that X2 has not significant relationship with X1 (Fig. 2.10). In other words, one of its explicit hypotheses isn't confirmed. Therefore, this path diagram isn't a fitted one for that data; this causal model isn't verified.

2.5 Identifying Various Effects

If a path diagram is a fitted one, we can examine various effects of the independent variables on the dependent variable. This final stage is the most important aspect of path analysis, because the *various effects* of the independent variables can be determined only by the path analysis. As mentioned previously, every independent variable can have five effects on the dependent variable. Indeed, every independent variable has three *actual effects* on the dependent variable: raw, direct and pure effects. In addition, every independent variable may have two *potential effects*: indirect and spurious effects.

2.5.1 Raw Effect

The *raw effect* of an independent variable on the dependent variable is the *bivariate correlation*[3] between them. In other words, the raw effect of

---

[3]The bivariate correlation (or Pearson correlation) is the regression coefficient in a linear regression model with only one independent variable, where both variables are expressed in the form of standard measures (the difference between a true value and the mean divided by standard deviation).

an independent variable is its effect on the dependent variable where no variable is controlled or held constant. The value of a raw effect indicates the change in the dependent variable by standard deviation for every increase of one standard deviation in the independent variable.

For example, in the path diagram of Example 2.1 (Fig. 2.1), the raw effect of X1 on the independent variable (X3) is 0.801 which means X3 increases 0.801 standard deviation for every increase of one standard deviation in X1 (Fig. 2.11).

In the linear regression analysis, the bivariate correlation is presented as *zero-order correlation* in the Coefficients output (Fig. 2.12).

> The raw effect of an independent variable is its effect on the dependent variable without controlling any variable.

Decomposing Raw Effect

The raw effect of an independent variable on the dependent variable decomposes into three components:

$$Raw = Direct + Indirect + Spurious \qquad (2.1)$$

Correlations

		X1	X2	X3
X1	Pearson Correlation	1	.646[**]	.801[**]
	Sig. (2-tailed)		.002	.000
	N	20	20	20
X2	Pearson Correlation	.646[**]	1	.955[**]
	Sig. (2-tailed)	.002		.000
	N	20	20	20
X3	Pearson Correlation	.801[**]	.955[**]	1
	Sig. (2-tailed)	.000	.000	
	N	20	20	20

**. Correlation is significant at the 0.01 level (2-tailed).

Fig. 2.11 Correlations of X1, X2 and X3, Example 2.1

Coefficients[a]

Model		Unstandardized Coefficients		Standardized Coefficients			Correlations		
		B	Std. Error	Beta	t	Sig.	Zero-order	Partial	Part
1	(Constant)	21.274	.537		39.609	.000			
	X1	4.724	.823	.317	5.741	.000	.801	.812	.242
	X2	1.034	.076	.750	13.606	.000	.955	.957	.573

a. Dependent Variable: X3

Fig. 2.12 Coefficients of the regression of X3 on X1 and X2 with Correlations, Example 2.1 [This output is produced by Part Correlation Command (see Sect. 2.6).]

Each of these components is a distinct effect and has a different meaning. Some consider this decomposition of the raw as the basic *theorem* of the path analysis (e.g. Mueller et al. 1977: 323). In the following, each of these effects is separately examined.

2.5.2   Direct Effect

The *direct effect* of an independent variable is its *immediate effect* on the dependent variable. The value of the direct effect is the *standardized regression coefficient*[4] of the independent variable in the regression of the dependent variable on the independent variables. The direct effect indicates the direct change in the dependent variable by standard deviation for every increase of one standard deviation in the independent variable while controlling for the effects of the other independent variables. It is shown by an arrow in path diagram and called *path coefficient*, too.

For example, in the path diagram of Fig. 2.13, the direct effect or the path coefficients are the standardized regression coefficients (Beta) in the regression of X3 on X1 and X2 (see Fig. 2.12). The path coefficient of X1 indicates that this independent variable has a direct effect of 0.317 on the dependent variable (X3). It means X3 increases directly 0.317 standard deviation for every increase of one standard deviation in X1 while controlling for the effects of X2 (the other independent variables in this path diagram). Moreover, X1 has a direct effect of 0.646 on X2, too (see Fig. 2.4).

> The direct effect of an independent variable is its effect on the dependent variable while controlling for the other independent variables.

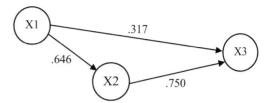

Fig. 2.13 Path diagram with coefficients, Example 2.1 ($R^2 = 0.970$)

2.5.3   Indirect Effect

The *indirect effect* of an independent variable is its effect on the dependent variable through the other independent variables. In other words, the indirect effect of an independent variable is its *mediated effect*. The value of the indirect effect is equal to the sum of the multiplication of the path coefficients in the indirect path from the independent variable to the dependent variable. It indicates the indirect change in the dependent variable by standard deviation for every increase of one standard deviation in the independent variable, while controlling for the effects of the other independent variables.

For example, the path diagram of Fig. 2.13 suggests that the independent variable X1 has an indirect effect on the dependent variable X3 through the independent variable X2. Thus, the value of this indirect effect is the path coefficient of X1 to X2 multiple by the path coefficient of X2 to X3:

$$0.646 \times 0.750 = 0.484.$$

It means X3 increases indirectly 0.485 standard deviation for every increase of one standard deviation in X1 while controlling for the effect of X2.

However, the independent variable X2 has not any indirect effect on the dependent variable X3. In other words, indirect effect of X2 on X3 is zero.

Given the Eq. 2.1 and that the spurious effect of an exogenous variable is zero (see following section), the value of the indirect effect of an exogenous variable can be calculated by this simple equation:

---

[4]Where the dependent variable and the independent variables are expressed in the form of standard measures which are also called standard scores or z-scores (the difference between a true value and the mean divided by standard deviation), the coefficients of the regression equation are called standardized regression coefficients (Beta).

$$\text{Indirect} = \text{Raw} - \text{Direct} \qquad (2.2)$$

For example, in the path diagram of Fig. 2.13, given the raw effect of the exogenous variable X1 is 0.801 (Fig. 2.12, column zero-order correlation) and its direct effect is 0.317, its indirect effect by this formula is:

$$\text{Indirect} = 0.801 - 0.317$$
$$= 0.484$$

Indirect effect an independent variable is its effect on the dependent variable which is mediated by the other independent variables in the path diagram.

2.5.4 Spurious Effect

The *spurious effect* of an independent variable on the dependent variable is a part of its effect which is due to the effect of the other variables in the path diagram. Thus, only the endogenous variables have spurious effect, because only this kind of variable is affected by the other independent variables in the path diagram. Therefore, the spurious effect of an exogenous variable is zero.

For example, in Fig. 2.13, only the endogenous variable X2 which is affected by the variable X1 has a spurious effect, while the exogenous variable X1 has not spurious effect.

The spurious effect is calculated by a complex formula. However, there is a simple way to obtain it. Since the raw effect of an independent variable is the sum of its direct, indirect and the spurious effect on the dependent variable (see Eq. 2.1), then:

$$\text{Spurious} = \text{Raw} - (\text{Direct} + \text{Indirect}) \qquad (2.3)$$

For example, in path diagram of Fig. 2.13, given the direct effect of X2 on X3 is 0.750, its indirect effect is 0.000 and its raw effect (zero-order correlation) is 0.955 (Fig. 2.12), then the spurious effect of the endogenous variable X2 on X3 is:

$$\text{Spurious} = 0.955 - (0.750 + 0.000)$$
$$= 0.205$$

The spurious effect of an endogenous variable is a part of its effect on the dependent variable that is due to the other independent variables.

2.5.5 Pure Effect

The *pure effect* of an independent variable is its effect on the dependent variable, after removing effects of the other independent variables affecting it. Indeed, the pure effect of an independent variable is the correlation between the dependent variable and the residual of the regression of that independent variable on the other independent variables which have causal effects on it. Thus, the *pure effect* in path analysis is the *unique effect* of an independent variable on the dependent variable.

Pure effect of an independent variable is its unique effect on the dependent variable.

In a path diagram, the pure effect of an exogenous variable is its raw effect (or the sum of its direct and indirect effects), because an exogenous variable is a variable that it isn't affected by any independent variable in the path diagram. However, the pure effect of an endogenous variable is its semipartial correlation in the linear regression of the dependent variable on it and the other independent variables which affect it.

In a multiple linear regression model, the *semipartial correlation* is the bivariate correlation between the dependent variable and an independent variable after the effects of the other independent variable are taken out of that independent variable. In SPSS, the semipartial correlation is present in the Coefficients output as Part Correlation.

For example, in the path diagram of Fig. 2.13, the pure effect of the exogenous variable X1 is its raw effect, which is 0.801 (Fig. 2.12). And, the pure effect of the endogenous variable X2 is its semipartial correlation in the regression of X3 on it and X1 (because X1 affects X2), which is 0.573 (see Fig. 2.12, column Part).

> An exogenous variable hasn't any spurious effect on the dependent variable.

As another example, in the path diagram of Fig. 2.14, the pure effect of the endogenous variable X2 is its semipartial correlation in the regression of X4 on it and X1 (because only X1 affects it), or the pure effect of the endogenous variable X3 is its semipartial correlation in the regression of X4 on the it, X1 and X2 (because both X1 and X2 affect it).

Squared Pure Effect

The *squared pure effect* of an independent variable is its *unique contribution* to the variation of the dependent variable. Indeed, the squared pure effect of an independent variable indicates the proportion of the variance in the dependent variable, which is exclusively due to that independent variable in a given path diagram.

Therefore, the sum of the squared pure effects is equal to the coefficient of determination (R^2), which indicates the proportion of the variance in the dependent variable that is due to the independent variables in the path diagram.

> The pure effect of an exogenous variable on the dependent variable is its raw effect.

For example, in the path diagram of Fig. 2.13, given that the pure effect of X1 is 0.801, and of X2 is 0.573, the sum of the squared pure effects is:

$$0.801^2 + 0.573^2 = 0.642 + 0.328 = 0.970$$

Which is equal to the coefficient of determination of that path diagram, $R^2 = 0.970$ (see Fig. 2.5).

2.6   Comparison of Effects

The path analysis is basically to identify and determine the various effects of the independent variables on the dependent variable. Thus, it allows comparing the various effects of the independent variables. This *comparison of effects* reveals that the direct effect of which independent variable is more or the indirect or spurious effect of which independent variable is more. Above all, the comparison of the pure effects shows what independent variable has largest unique effect on the dependent variable and especially squared pure effects show that the unique contribution of what independent variable is more.

To facilitate the comparison of effects of the independent variables on the dependent variable, it is better to sum up them on a table which can be named effect matrix, such as Table 2.4.

Effect Matrix

The *effect matrix* is a table which includes the various effects of all independent variables on the dependent variable in order to compare the effects. In an effect matrix, such as Table 2.4, the first column is the list of the independent variables.

> The pure effect of an endogenous variable is its semipartial correlation with the dependent variable.

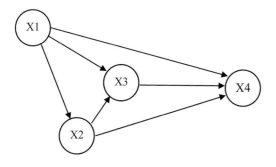

Fig. 2.14 Path diagram, Example 2.4

Table 2.4 Effect matrix of the path diagram of Fig. 2.13 ($R^2 = 0.970$)

Independent variable	Raw	Direct	Indirect	Spurious	Pure	Squared pure
X1	0.801	0.317	0.484[a]	0.000	0.801	0.642
X2	0.955	0.750	0.000	0.205	0.573	0.328
Total	-	–	–	–	–	0.970

[a]$0.646 \times 0.750 = 0.484$ (see Fig. 2.13)

For example, the effect matrix of the path diagram of Fig. 2.13 (Table 2.4) shows that the direct effect of X2 is more than the double of X1 (0.750/0.317 = 2.37). However, X1 has an indirect effect of 0.484, while X2 has not any indirect effect. Also, X1 has not spurious effect, while X2 has a spurious effect of 0.205. Thus, the pure effect of X1 is more than of X2 and its squared pure effect is near to the double of X2 (0.642/0.328 = 1.96).

Therefore, in spite of that the direct effect of X2 is much more than of X1, the contribution of the independent variable X1 to the variation of the dependent variable X3 is much more.

In the next example, all stages of a four-variable path analysis are carried out.

> The pure effect of an endogenous variable is its effect on the dependent variable after removing effects of the other independent variables affecting it.

Example 2.4

Based on a theoretical framework, a researcher assumes three variables of gender (X1), education (X2) and post (X3) are important factors of salary (X4) of the employees. His argument is briefly summarized as follows: In the contemporary societies, the general status of men is higher than women in many societies. Hence, the average income of men is greater than women. Also, education and specialization is an important component of modern economy. Therefore, the salary has a direct relationship with education; the more education, the more salary. The administrative hierarchy is also an important component of organizations and as a result, the post of employees has a direct relationship with

salary; the higher post, the more salary. In addition, due to the high status of men, the average education of men is greater than women; gender (X1) affects education (X2). Also, for this reason and the importance of education, the high organizational posts are usually assigned to men and higher educated individuals. In other words, gender (X1) and education (X2) affect the organizational post (X3). This causal chin is presented as a path diagram in Fig. 2.14. ◄

Also, suppose the appropriate data to test the path diagram of Fig. 2.14 are collected from the survey of a probability sample of people in a town (Table 2.5).

As mentioned previously, the first step of path analysis is the examination of the technical assumptions.

Examination of Technical Assumptions

1. Gender (X1) is a binary variable, coded as 0 and 1 (0 for female and 1 for male). Education (X2) is a quantitative variable (by years of schooling). Post is considered as a quantitative variable from 0 to 4. And, salary (X4) is obviously a quantitative variable.

2. The relationship between the dependent variable and each quantitative independent variable is linear" as shown in Figs. 2.15 and 2.16. The additive relationship and one-way flow path is also granted.

> The squared pure effect of an independent variable indicates the proportion of the variance in the dependent variable which is exclusively due to it.

3. According to third assumption, "the observations are independent of each other", the adjacent residuals should not be

Table 2.5 Distribution of gender, education, post and salary, Example 2.4

No.	X1 Gender	X2 Education	X3 Post	X4 Salary
1	0	0	0	2000
2	1	5	5	7450
3	1	4	4	6320
4	0	1	1	3130
5	0	1	1	3130
6	1	3	2	5640
7	0	2	2	3460
8	0	3	1	3490
9	0	1	1	3530
10	0	1	1	3530
11	1	1	1	3730
12	0	2	2	3860
13	0	3	2	4240
14	0	2	2	4260
15	0	4	1	4270
16	0	4	1	4270
17	1	2	2	4460
18	1	2	2	4460
19	0	4	1	4670
20	1	3	2	4840
21	1	2	2	4860
22	1	2	2	4860
23	0	3	3	4990
24	0	4	2	5020
25	1	3	2	5240
26	1	4	3	5570
27	1	4	3	5570
28	1	3	3	5590
29	1	3	2	5640
30	0	3	0	3140
31	1	3	3	5990
32	1	3	3	5990
33	1	4	4	6320
34	0	1	1	2730
35	0	0	0	2000
36	1	5	4	7100
Total	–	95	71	165,350

correlated with each other. As it is evident from the last column of the Model Summary of the regression of the dependent variable X4 on the three independent variables of X1 to X3 (Fig. 2.17), Durbin-Watson Test is 2.021, which is almost equal to 2. Therefore, it indicates the lack of correlation between observations.

4. According to fourth assumption, "the distribution of the dependent variable for each value of the independent variable is normal". As Fig. 2.18 shows, the histogram of the distribution of the standardized residuals in the linear regression of the dependent variable X4 on the three independent variables X1 to X3 is close to the normal distribution. Therefore, assuming normal distribution of residuals isn't questioned.

5. And fifth assumption says the variance of the dependent variable should be the same for all values of the independent variables. The scatterplot of regression standardized residual and regression standardized predicted value (Fig. 2.19) indicates that the spread of standardized residual doesn't increase or decrease with the magnitude of predicted value; the standardized residuals is randomly scattered around a horizontal line. In other words, there isn't any relationship between residuals and predicted values. Thus, the assumption of equal variances isn't questioned.

Now, after examining the technical assumptions of the path analysis and ensuring that there isn't severe violation of them, it turns to test the hypotheses of the path diagram.

Test of Path Diagram
To test the path diagram of Fig. 2.14 with two endogenous variables (X2 and X3) requires three linear regression equations:

1. The regression of the dependent variable X4 (salary) on three independent variables X1 (gender), X2 (education) and X3 (post). The output of this regression indicates that the dependent variable has significant

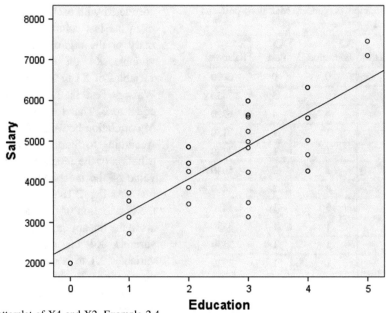

Fig. 2.15 Scatterplot of X4 and X2, Example 2.4

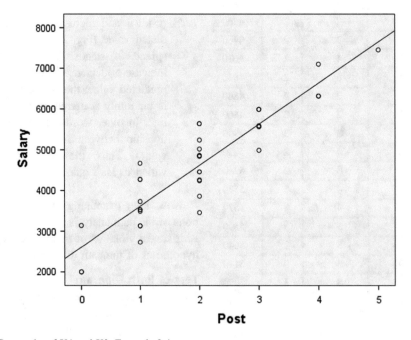

Fig. 2.16 Scatterplot of X4 and X3, Example 2.4

Model Summary[b]

Model	R	R Square	Adjusted R Square	Std. Error of the Estimate	Durbin-Watson
1	.975[a]	.950	.945	307.088	2.021

a. Predictors: (Constant), X3 Post, X1 Gender, X2 Education

b. Dependent Variable: X4 Salary

Fig. 2.17 Model summary of the regression of X4 on X1, X2 and X3, Example 2.4

Fig. 2.18 Histogram of standard residuals, Example 2.4

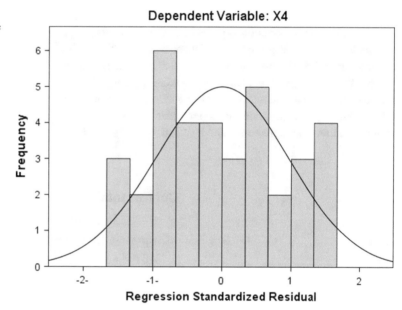

Fig. 2.19 Scatterplot of regression standardized residual and regression standardized predicted value, Example 2.4

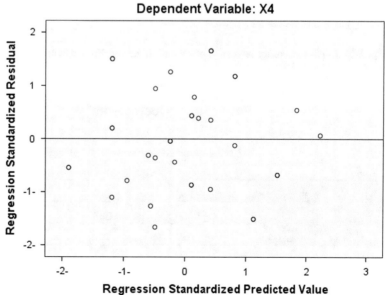

relationships with all independent variables (Fig. 2.20).

2. The regression of endogenous variable X3 (post) on X1 (gender) and X2 (education). It indicates that X3 has significant relationship with X1 and X2 (Fig. 2.21).

3. The regression of endogenous variable X2 on X1, which shows that X2 (education) has significant relationship with X1 (Fig. 2.22).

Also, all observed relationships are positive, which mean they correspond to the theoretical expectations: the average income of men is greater than women; the salary has a direct relationship with education; the post has a direct relationship with salary; the posts are mainly assigned to men and higher educated individuals;

and the education of men is averagely greater than of women.

Moreover, this path diagram has not any implicit hypothesis. Thus, all hypotheses of the path diagram of Fig. 2.14 are confirmed. Therefore, the path diagram is a fitted one (at least with data of Table 2.5).

The fitness index of path diagram, the coefficient of determination of first multiple linear regression is 0.950 (Fig. 2.17). It means most (95%) of the variation of salary is explained by gender, education and post. In causal terms, it means 95% of the difference in paid employees' salaries is due to their gender, education and post.

Thus, we can identify the effects of the independent variables of the path diagram of Fig. 2.14.

Coefficients[a]

Model		Unstandardized Coefficients		Standardized Coefficients		
		B	Std. Error	Beta	t	Sig.
1	(Constant)	2167.080	118.291		18.320	.000
	X1 Gender	756.878	134.658	.293	5.621	.000
	X2 Education	404.923	53.943	.405	7.507	.000
	X3 Post	496.389	72.775	.448	6.821	.000

a. Dependent Variable: X4 Salary

Fig. 2.20 Coefficients of the regression of X4 on X1, X2 and X3, Example 2.4

Coefficients[a]

Model		Unstandardized Coefficients		Standardized Coefficients		
		B	Std. Error	Beta	t	Sig.
1	(Constant)	.232	.280		.829	.413
	X1 Gender	1.068	.263	.458	4.063	.000
	X2 Education	.457	.102	.507	4.498	.000

a. Dependent Variable: X3 Post

Fig. 2.21 Coefficients of the regression of X3 on X1 and X2, Example 2.4

Coefficients[a]

Model		Unstandardized Coefficients		Standardized Coefficients		
		B	Std. Error	Beta	t	Sig.
1	(Constant)	2.167	.292		7.413	.000
	X1 Gender	.944	.413	.365	2.285	.029

a. Dependent Variable: X2 Education

Fig. 2.22   Coefficients of the regression of X2 on X1, Example 2.4

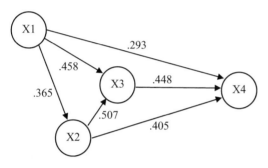

Fig. 2.23   Path diagram with path coefficients, Example 2.4, ($R^2 = 0.950$)

Effects of Independent Variables

First, the path coefficients are introduced in the path diagram (Fig. 2.23). This path coefficients are obtained from the regression of X4 on X1 to X3 (Fig. 2.20), the regression of X3 on X1 and X2 (Fig. 2.21) and the regression of X2 on X1 (Fig. 2.22).

Now, we can constitute the effect matrix of this path diagram (Table 2.6).

Raw Effects   In the effect matrix of Example 2.4, the raw effects which are the bivariate correlations between the independent variables and

the dependent variable are derived from Fig. 2.24 (zero-order correlation).

Direct Effects   The direct effects which are the standardized regression coefficients of the regression of X4 on X1 to X3 (Fig. 2.24) are the path coefficients in the path diagram of Fig. 2.23.

Indirect Effects   The indirect effect of the independent variable X1 on the dependent variable X4 is by three way: (1) from X2 to X4; (2) from X2 to X3 to X4; and (3) from X3 to X4. The value of first path is:

$$0.365 \times 0.405 = 0.148.$$

The value of second path is:

$$0.365 \times 0.507 \times 0.448 = 0.083.$$

And the value of third path is:

$$0.458 \times 0.448 = 0.205.$$

Thus, the indirect effect of X1 is the sum of these three values:

Table 2.6   Effect matrix of path diagram, Example 2.4 ($R^2 = 0.950$)

Independent variable	Raw	Direct	Indirect	Spurious	Pure	Squared pure
X1 (Gender)	0.728	0.293	0.435	0.000	0.728	0.530
X2 (Education)	0.814	0.405	0.227	0.182	0.589	0.347
X3 (Post)	0.909	0.448	0.000	0.461	0.270	0.073
Total	–	–	–	–	–	0.950

Coefficients[a]

Model		B	Std. Error	Beta	t	Sig.	Zero-order	Partial	Part
		Unstandardized Coefficients		Standardized Coefficients			Correlations		
1	(Constant)	2167.080	118.291		18.320	.000			
	X1 Gender	756.878	134.658	.293	5.621	.000	.728	.705	.222
	X2 Education	404.923	53.943	.405	7.507	.000	.814	.799	.297
	X3 Post	496.389	72.775	.448	6.821	.000	.909	.770	.270

a. Dependent Variable: X4 Salary

Fig. 2.24 Coefficients of the regression of X4 on X1, X2 and X3 with correlations, Example 2.4

$$0.148 + 0.083 + 0.205 = 0.436.$$

The indirect effect of this exogenous variable could be calculated by the simple Eq. 2.2, too:

$$\begin{aligned} \text{Indirect} &= \text{Raw} - \text{Direct} \\ &= 0.728 - 0.293 \\ &= 0.435 \end{aligned}$$

The slight difference between the two values is due to the rounding error.

The indirect effect of the independent variable X2 is from X3 to X4:

$$0.507 \times 0.448 = 0.227.$$

And, the independent variable X3 has not any indirect effect on X4.

Spurious Effects The spurious effect of the independent variable X1, which isn't affected by X2 and X3, is zero. However, X2, which is affected by X1, has a spurious effect of:

$$\begin{aligned} \text{Spurious} &= \text{Raw} - (\text{Direct} + \text{Indirect}) \\ &= 0.814 - (0.405 + 0.227) \\ &= 0.182 \end{aligned}$$

Also, X3, which is affected by X1 and X2 and has not indirect effect, has a spurious effect of:

$$\begin{aligned} \text{Spurious} &= \text{Raw} - (\text{Direct} + \text{Indirect}) \\ &= 0.909 - (0.448 + 0.000) \\ &= 0.182 \end{aligned}$$

Pure Effects The pure effect of the independent variable X1 is equal to its raw effect (0.728), because it isn't affected by X2 and X3; it is an exogenous variable. However, the pure effect of the endogenous variable X2 which is affected by X1, is its semipartial correlation in the regression of X4 on it and X1; it is 0.589 (last column of Fig. 2.25, Part Correlation).[5] The pure effect of X3 which is affected by X1 and X2, is its semipartial correlation in the regression of X4 on it, X1 and X2, which is 0.270 (Fig. 2.24, Part Correlation).

Comparison

The effect matrix of the path diagram of Example 2.4 (Table 2.6) shows that the direct effect of post (0.909) on salary is greater than two other independent variables. However, gender with an indirect effect of 0.435, has greater indirect effect, while post has not any indirect effect. Also, gender has not spurious effect, while post has a great spurious effect of 0.461.

---

[5]Why isn't the part correlation of X2 in Fig. 2.24 its pure effect?

Coefficients[a]

Model		Unstandardized Coefficients		Standardized Coefficients			Correlations		
		B	Std. Error	Beta	t	Sig.	Zero-order	Partial	Part
1	(Constant)	2282.312	180.602		12.637	.000			
	X1 Gender	1287.233	169.593	.498	7.590	.000	.728	.797	.463
	X2 Education	631.753	65.516	.632	9.643	.000	.814	.859	.589

a. Dependent Variable: X4 Salary

Fig. 2.25 Coefficients of the regression of X4 on X1 and X2 with correlations, Example 2.4

Thus, while the raw effect of post (0.909) is greater than of gender (0.728), the pure effect of gender (0.728) is greater than of post (0.270). The X1 has the most pure effect and the X3 has the least effect on X4. Indeed, the contribution of gender to the variation of salary is very high: its squared pure effect is 0.530, which is over 7 times more than squared pure effect of post $(0.530/0.073 = 7.26)$. Even the squared pure effect of education is near 5 times more than the squared pure effect of post $(0.347/0.073 = 4.75)$.

In this way, it is overt despite that X3 has the largest direct effect on X4. Its contribution to X4 is very little in comparison with contribution of X1, while X1 with lowest direct effect has very large contribution to the variation of X4.

This example used a hypothetical data, but the next example uses a relatively similar causal model with true data.

Example 2.5

Let's assume individuals who have higher social position earn more. As a result, individuals who are male, white, with higher education and occupational prestige earn more. In other words, gender (X1), race (X2), education (X3) and occupational prestige (X4) have causal effects on income (X5). Moreover, individuals who are male and white have higher education, too. Also, individuals who are male, white and with higher education have more high level occupations. This causal chin is presented as a path diagram presented in Fig. 2.26. ◄

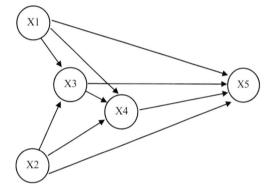

Fig. 2.26 Path diagram Example 2.5

Here, we test these hypotheses by using the data of the General Social Survey 1972[6] which is a probability sample of US people, as mentioned previously. In this file, we recoded SEX (Respondents sex) into a different variable, named gender (X1), coded as female = 0 and male = 1; RACE (Race of respondent) into a different variable, as race (X2), coded as black = 0 and white = 1. Also, we named EDUC (Highest year of school completed) as education (X3), and PRESTIGE (R's occupational prestige score 1970) as prestige (X4). Also, we took the CONINC (R's Family income in constant dollars) as indicator of the income[7] (X5).

---

[6]http://www.gss.norc.org/get-the-data/spss.

[7]We used the family income, because this survey did not include respondent income. But, since in this analysis the occupational prestige indicates that respondent is employed, the respondent income is either one of the

Now, in the first step of the path analysis, the technical assumptions are examined.

Examination of Technical Assumptions

1. Gender and race are binary variables, coded as 0 and 1, as mentioned previously. Education (by years of schooling), prestige (by score) and income (by dollars) are quantitative variables.
2. The relationship between the dependent variable and the quantitative independent variables are rather "linear" as shown in Figs. 2.27 and 2.28. The additive relationship and one-way flow path is also granted.
3. According to third assumption, "the observations are independent of each other", the adjacent residuals should not be correlated with each other. As it is evident from the last column of the Model Summary of the regression of the dependent variable X5 on the four independent variables of X1 to X4 (Fig. 2.29), Durbin-Watson Test is 1.565, which is near to 2 than to 0 or 4. Therefore, it indicates the lack of correlation between observations.
4. According to fourth assumption, "the distribution of the dependent variable for each value of the independent variable is normal". As Fig. 2.30 indicates, the histogram of the distribution of the standardized residuals in the linear regression of the dependent variable X5 on the four independent variables X1 to X4 is close to the normal distribution. Therefore, assuming normal distribution of residuals isn't questioned.
5. And fifth assumption says the variance of the dependent variable should be the same for all values of the independent variables. The scatterplot of regression standardized residual and regression standardized predicted value (Fig. 2.31) shows that the spread of standardized residual doesn't increase or decrease with the magnitude of

predicted value; the standardized residuals is randomly scattered around a horizontal line. In other words, there isn't any relationship between residuals and predicted values. Thus, the assumption of equal variances isn't questioned.

Now, after examining the technical assumptions of the path analysis and ensuring that there isn't severe violation of them, it turns to test the hypotheses of the path diagram.

Test of Path Diagram

To test the path diagram of Fig. 2.26 with two endogenous variables (X3 and X4) requires three linear regression equations:

1. The regression of the dependent variable X5 (income) on four independent variables X1 (gender), X2 (race), X3 (education) and X4 (occupational prestige). The output of this regression indicates that the dependent variable has significant relationship with all these independent variables (Fig. 2.32).
2. The regression of the endogenous variable X4 (occupational prestige) on three other independent variable X1 (gender), X2 (race) and X3 (education). It indicates that X4 has significant relationship with them (Fig. 2.33).
3. The regression of the endogenous variable X3 (education) on two independent variable X1 (gender) and X2 (race). It shows that X3 has significant relationship with them (Fig. 2.34).

In sum, all coefficients are significant at level less than 0.05. In addition, all observed relationships are positive, which mean they correspond to the theoretical expectations: the individuals who are male, white, and have higher education and higher occupational prestige earn more. Also, men and white have higher education, too. Moreover, men, white and individuals with higher education have occupations of higher prestige.

Also, as Fig. 2.35 shows, there isn't any correlation between X1 and X2. Therefore, the

---

main components of the family income or the only component. Thus, the latter is a good indicator of the former.

Fig. 2.27 Scatterplot of X5 and X3, Example 2.5

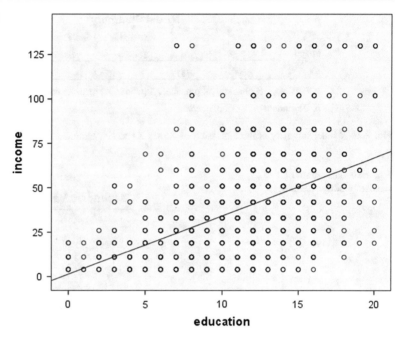

Fig. 2.28 Scatterplot of X5 and X4, Example 2.5

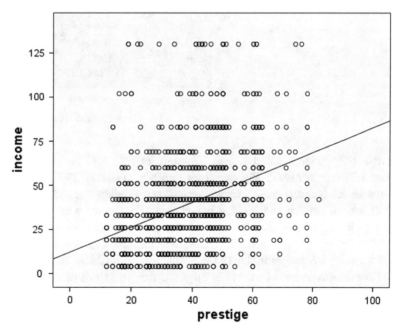

Model Summary^b

Model	R	R Square	Adjusted R Square	Std. Error of the Estimate	Durbin-Watson
1	.480^a	.231	.228	23.201	1.565

a. Predictors: (Constant), X4 prestige, X1 gender, X2 race, X3 education

b. Dependent Variable: X5 income

Fig. 2.29 Model summary of the regression of X5 on X1 to X4, Example 2.5

Fig. 2.30 Histogram of standard residuals, Example 2.5

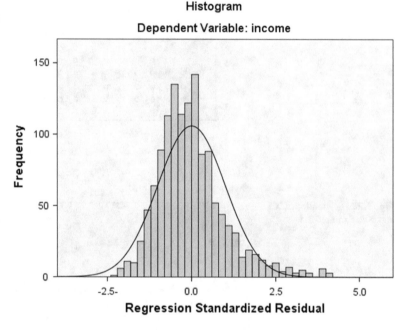

implicit hypothesis of the path diagram of Fig. 2.26 that there isn't any causal relationship between X1 (gender) and X2 (race) is confirmed.

Thus, all explicit and implicit hypotheses of the path diagram of Fig. 2.26 are confirmed. Therefore, this path diagram is a fitted one (at least among US people in the 1972).[8]

The fitness index of this path diagram, the coefficient of determination of first multiple linear regression is 0.231 (Fig. 2.36), which means near one fourth of the variation of income is explained by gender, race, education and occupational prestige. In causal terms, it means about

23% of the difference in the income of Americans in 1972 was due to their gender, race, education and occupational prestige.

But, what variable is most important in the variation of income? To answer this question, we present the various effects of the independent variables on income as an effect matrix and compare them.

Effects of Independent Variables

First, the path coefficients are introduced in path diagram (Fig. 2.37). This path coefficients are obtained from the regression of X5 on X1 to X4 (Fig. 2.32), the regression of X4 on X1 to X3 (Fig. 2.33) and the regression of X3 on X1 and X2 (Fig. 2.34). Then, we can constitute the effect matrix of this path diagram (Table 2.7).

---

[8]Do you think this causal model (path diagram) is a fitted causal model nearly half a century later? You can test your hypotheses by the data of the General Social Survey 2018 (see Exercise 2.6).

Fig. 2.31 Scatterplot of regression standardized residual and regression standardized predicted value, Example 2.5

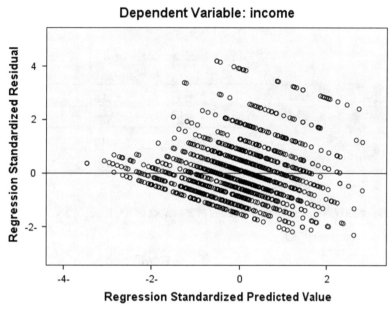

Coefficients[a]

Model		Unstandardized Coefficients		Standardized Coefficients		
		B	Std. Error	Beta	t	Sig.
1	(Constant)	-9749.035-	2572.336		-3.790-	.000
	X1 gender	5637.819	1293.976	.106	4.357	.000
	X2 race	8119.483	1822.858	.112	4.454	.000
	X3 education	2506.233	222.162	.330	11.281	.000
	X4 prestige	271.016	58.354	.139	4.644	.000

a. Dependent Variable: X5 income

Fig. 2.32 Coefficients of the regression of X5 on X1 to X4, Example 2.5

Raw Effects In the effect matrix of Example 2.5 (Table 2.7) the raw effects which are the bivariate correlations between the independent variables and the dependent variable are derived from Fig. 2.38 (zero-order correlation).

Direct Effects The direct effects which are the standardized regression coefficients of the regression of X5 on X1 to X4 (Fig. 2.38, Beta) are the path coefficients in the path diagram of Fig. 2.37.

Indirect Effects The indirect effect of the independent variable X1 on the dependent variable X5 is by three path: (1) from X3 to X5; (2) from X3 to X4 to X5; and from X4 to X5. As mentioned previously, the sum of the indirect effects of an exogenous variable such as X1 is simply the difference between its raw effect and direct effect (Eq. 2.2). Thus, the indirect effect of X1 is:

Coefficients[a]

| Model | | Unstandardized Coefficients | | Standardized Coefficients | | |
		B	Std. Error	Beta	t	Sig.
1	(Constant)	9.422	1.150		8.191	.000
	X1 gender	1.892	.585	.070	3.235	.001
	X2 race	5.894	.815	.160	7.230	.000
	X3 education	2.019	.086	.516	23.364	.000

a. Dependent Variable: X4 prestige

Fig. 2.33 Coefficients of the regression of X4 on X1 to X3, Example 2.5

Coefficients[a]

| Model | | Unstandardized Coefficients | | Standardized Coefficients | | |
		B	Std. Error	Beta	t	Sig.
1	(Constant)	9.493	.227		41.792	.000
	X1 gender	.417	.169	.060	2.470	.014
	X2 race	1.936	.230	.205	8.410	.000

a. Dependent Variable: X3 education

Fig. 2.34 Coefficients of the regression of X3 on X1 and X2, Example 2.5

Fig. 2.35 Correlations of X1 and X2, Example 2.5

Correlations

		X1 gender	X2 race
X1 gender	Pearson Correlation	1	-.002-
	Sig. (2-tailed)		.937
	N	1613	1609
X2 race	Pearson Correlation	-.002-	1
	Sig. (2-tailed)	.937	
	N	1609	1609

Model Summary

Model	R	R Square	Adjusted R Square	Std. Error of the Estimate
1	.480[a]	.231	.228	23302.580

a. Predictors: (Constant), X4 prestige, X1 gender, X2 race, X3 education

Fig. 2.36 Model summary of the regression of X5 on X1 to X4, Example 2.5

Table 2.7 Effect matrix of path diagram, Example 2.5 (R^2 = 0.231)

Independent variable	Raw	Direct	Indirect	Spurious	Pure	Squared pure
X1 Gender	0.122	0.106	0.016	0.000	0.122	0.015
X2 Race	0.224	0.112	0.112	0.000	0.224	0.050
X3 Education	0.432	0.330	0.072	0.030	0.390	0.152
X4 Occupation prestige	0.364	0.139	0.000	0.225	0.112	0.013
Total	–	–	–	–	–	0.230

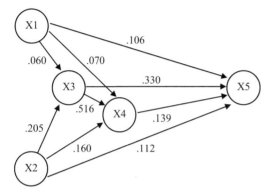

Fig. 2.37 Path diagram with coefficients, Example 2.5, (R^2 = 0.231)

$$\text{Indirect} = \text{Raw} - \text{Direct}$$
$$= 0.122 - 0.106$$
$$= 0.016$$

The indirect effect of X2 (other exogenous variable) is:

$$\text{Indirect} = \text{Raw} - \text{Direct}$$
$$= 0.224 - 0.112$$
$$= 0.112$$

The indirect effect of the independent variable X3 is from X4 to X5. As a result, it is:

$$0.516 \times 0.139 = 0.072$$

And, the independent variable X4 has not any indirect effect on X5.

Spurious Effects The spurious effects of the independent variable X1 and X2 which are the exogenous variables (i.e. aren't affected by the other independent variables) are zero.

However, X3 which is affected by X1 and X2 has a spurious effect of (Eq. 2.3):

$$\text{Spurious} = \text{Raw} - (\text{Direct} + \text{Indirect})$$
$$= 0.432 - (0.330 + 0.072)$$
$$= 0.030$$

And, X4, which is affected by X1 to X3, has a spurious effect of:

$$\text{Spurious} = \text{Raw} - (\text{Direct} + \text{Indirect})$$
$$= 0.364 - (0.139 + 0.000)$$
$$= 0.225$$

Pure Effects The pure effect of X1 is equal to its raw effect (0.122), because it isn't affected by the other independent variables (it is an exogenous variable). Also, the pure effect of the other exogenous variable X2 is its raw effect (0.224).

However, the pure effect of the endogenous variable X3, which is affected by X1 and X2, is its semipartial correlation in the regression of X5 on it and both X1 and X2. As Fig. 2.39 shows, it is 0.390 (last column, Part Correlation).[9] And the pure effect of the other endogenous variable (X4) which is affected by X1 to X3 is its semipartial correlation in the regression of X5 on it and X1 to X3. It is 0.112 (Fig. 2.38, Part Correlation).

Comparison

The effect matrix of path diagram Example 2.5 (Table 2.7) shows that education (X3) has the largest direct effect as well as the largest pure effect on income. Indeed, its unique contribution

---

[9]Why isn't the part correlation of X2 in Fig. 2.38 its pure effect?

Coefficients[a]

Model		Unstandardized Coefficients B	Unstandardized Coefficients Std. Error	Standardized Coefficients Beta	t	Sig.	Correlations Zero-order	Correlations Partial	Correlations Part
1	(Constant)	-9749.035-	2572.336		-3.790-	.000			
	X1 gender	5637.819	1293.976	.106	4.357	.000	.122	.119	.105
	X2 race	8119.483	1822.858	.112	4.454	.000	.224	.122	.107
	X3 education	2506.233	222.162	.330	11.281	.000	.432	.296	.272
	X4 prestige	271.016	58.354	.139	4.644	.000	.364	.127	.112

a. Dependent Variable: X5 income

Fig. 2.38 Coefficients of the regression of X5 on X1 to X4, with correlations, Example 2.5

Coefficients[a]

Model		Unstandardized Coefficients B	Unstandardized Coefficients Std. Error	Standardized Coefficients Beta	t	Sig.	Correlations Zero-order	Correlations Partial	Correlations Part
1	(Constant)	-7601.438-	2385.024		-3.187-	.001			
	X1 gender	6208.475	1230.478	.117	5.046	.000	.137	.131	.117
	X2 race	10112.855	1710.700	.140	5.912	.000	.222	.153	.137
	X3 education	3043.284	180.514	.398	16.859	.000	.432	.403	.390

a. Dependent Variable: X5 income

Fig. 2.39 Coefficients of the regression of X5 on X1 to X3 with correlations, Example 2.5

to the variation of income (squared pure effect) is over 3 times more than one of race (0.152/0.050 = 3.04), over 10 times more than one of gender (0.152/0.015 = 10.13) and near 12 times more than one of the occupational prestige (0.152/0.013 = 11.69). Thus, it is a very important factor in the income.

Moreover, the direct effect of the occupational prestige is larger than the one of gender and race, while its pure effect is less than their pure effects.

Thus, near half of a century ago, the unique contribution of education as a modern achieved status to the variation of income was over 2 times more than the sum of the contributions of both gender and race, which are two important traditional ascribed status (0.152/[0.050 + 0.015] = 2.39).[10]

2.6.1 Importance of Independent Variables

Determining the various effects of the independent variables on the dependent variable is as

[10]Do you think these relationships are also true near half a century later? (see Exercise 2.7).

Fig. 2.40 Data editor, analyze, regression, linear…

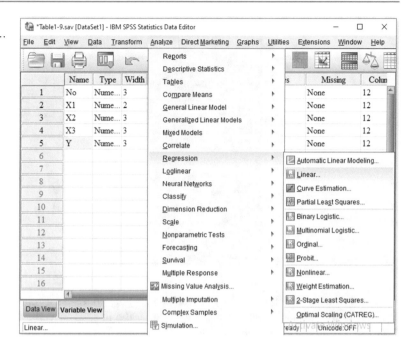

Fig. 2.41 Linear regression dialog box

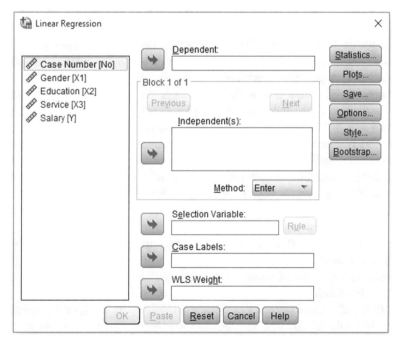

important as testing the assumed causal relationship. Path analysis performs this task by identifying the various effects of independent variables. In this way, it allows determining the importance of independent variables by each of the distinct effects.

Unfortunately, there are few books that address this issue. These books marginally point

Fig. 2.42 Linear regression: statistics dialogue box

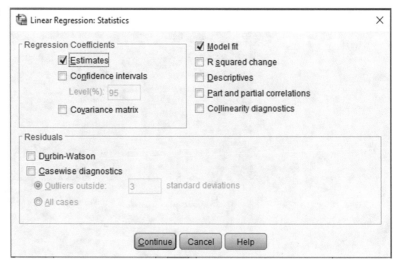

to it and usually confuse it. As mentioned in Preface, in the statistics books (e.g. Foster et al. 2006) or research methods books (e.g. de Vaus 2002) the discussion of the effects of the independent variables on the dependent variable is usually restricted to direct and indirect effects. And the sum of these two effects as total effect is implicitly seen as pure effect. Sometimes, the comparison is limited only to compare the standardized regression coefficient and again implicitly it is considered as the indicators of the contribution of the independent variables to the variation of the dependent variable (e.g. Allison 1999). It generally point out that the standardized regression coefficient is the most used measure of relative importance or size of the effect of a variable (e.g. see Darlington and Hayes 2017: 269).

However, the examples in this chapter show that, contrary to this common confusion, the importance of independent variables differs in terms of their various effects. A variable that is less important in one respect may be more important in other respect.

Example 2.4 showed that X3 has the very larger direct effect in comparison with X1, but its

pure effect and its contribution to the variation of the dependent variable is very little than of X1. Or, Example 2.5 showed while the direct effect of the occupational prestige is larger than of gender and race, its pure effect is less than those independent variables.

2.7 Part Correlation Command

To obtain a part correlation in SPSS:
1. In the Data Editor (data file) as shown in Fig. 2.40, select **Analyze → Regression → Linear...**,
2. In the **Linear Regression** dialogue box (Fig. 2.41), select the dependent variable from the list of variables and transfer it into, the **Dependent:** box, and the independent variables into the **Independent(s):** box,
3. Click on the **Statistics...** button,
4. In the **Linear Regression: Statistics** dialogue box (Fig. 2.42), click on the **Part partial correlation**,
5. Click on the **Continue** button,
6. In the **Linear Regression** dialogue box, click **OK**.

2.8 Summary

This chapter showed that the path analysis which is based on a set of multiple linear regressions is a powerful technique to test not only the causal relationships between a set of independent variables and the dependent variable, but also the relationships between the independent variables. It is shown that a causal model which is a theoretical model can be expressed as a path diagram that presents assumed causal relationships between variables. A path analysis involves three stages: the examination of technical assumptions, the test of path diagram, and the presentation of various effects. It is illustrated how an independent variable has three actual effects on the dependent variable (raw, direct and pure effects) and two potential (indirect and spurious) effects. The comparison of the various effects of independent variables shows that, contrary most texts, the pure effect isn't always the sum of the direct and indirect effects. Indeed it is only an exogenous variable that its pure effect is the sum of its direct and indirect effects. But, the pure effect of an endogenous variable is its effect after removing effects of the other independent variables affecting it. So, the pure effect of an endogenous variable is always smaller than the sum of its direct and indirect effects. Finally, it can be specified the unique contribution of each independent variable on the variation of the dependent variable.

2.9 Exercises

2.1 A researcher assumed that three variables X1, X2 and X3 have causal effects on the dependent variable (X4); two independent variables X1 and X2 have causal effects on the X3; and the independent variable X1 has causal effect on X2. Another researcher also assumed that three variables X1, X2 and X3 have causal effects on the dependent variable (X4); two independent variables X1 and X2 have causal effects on the X3. However, she or he did not assume any causal relationship between X1 and X2.

Draw a path diagram for each of two causal models.

2.2 Given the data of Table 2.8, which of above path diagrams (causal models) is a fitted model? (Check the technical assumptions of path analysis and hypotheses of path diagram).

2.3 Interpret the coefficient of determination of the fitted model.

2.4 Complete the effect matrix of the fitted model.

2.5 Compare and interpret the various effects of the independent variables.

2.6 Test the hypotheses of the path diagram of Fig. 2.31 (Example 2.5) by using the data of the General Social Survey 2018[11] which is a probability sample of US people.[12] Is that path diagram a fitted one? Why?

2.7 How many fitted path diagram can be obtained by the independent variables of Exercise 2.6? What is the most fitted one? Among these independent variables, the unique contribution of what variable on the variation of income is biggest?

2.8 Based on a theoretical arguments, a researcher assumed that four independent variables X1 (gender), X2 (SES: socioeconomic status of parents), X3 (education) and X4 (participation: membership in the social groups) have causal effects on X5 (happiness); three variables X1 to X3 have causal effects on X4; two variables X1 and X2 have causal effects on X3; but there isn't any causal relationship between X1 and X2. Draw a path diagram for above causal model.

2.9 Given the data of Table 2.9, is above path diagram a fitted model?

2.10 How much is the coefficient of determination of that model?

2.11 Complete the effect matrix of the causal model of Exercise 2.8.

---

[11]http://www.gss.norc.org/get-the-data/spss.

[12]In this file, the variables for that path diagram are: SEX (Respondents sex), RACE (Race of respondent), EDUC (Highest year of school completed), PRESTG10 (R's occupational prestige score) and CONINC (Family income in constant dollars).

Table 2.8 Distribution of X1 to X4, in a probability sample of people in a town

No.	X1	X2	X3	X4
1	5	2	12	23
2	1	3	11	16
3	5	1	9	18
4	5	3	9	20
5	0	1	0	1
6	4	1	8	16
7	1	2	8	10
8	3	2	10	18
9	4	2	11	21
10	2	2	9	16
11	1	1	4	5
12	0	3	5	10
13	2	3	12	21
14	3	1	7	13
15	3	3	13	25
16	0	2	6	11
17	4	3	12	26
18	2	1	6	15
19	0	3	5	10
20	1	3	11	16
21	2	3	12	21
22	0	2	6	11
23	2	2	9	16
24	4	2	11	21
25	0	1	0	1
26	2	1	6	15
27	4	1	8	16
28	5	3	9	20
29	3	3	13	25
30	3	2	10	18
31	1	1	4	7
32	5	1	9	18
33	4	3	12	26
34	5	2	12	23
35	1	2	8	10
36	3	1	7	15
Total	90	72	304	574

Table 2.9 Distribution of X1 to X5 in a probability sample of people in a town

No.	X1 Gender[a]	X2 SES	X3 Education	X4 Participation	X5 Happiness
1	0	0	0	0	1
2	0	0	1	1	1
3	0	0	1	1	1
4	0	1	1	1	2
5	0	1	2	1	2
6	0	1	2	2	2
7	0	2	1	1	2
8	0	2	1	1	2
9	0	2	3	2	2
10	0	3	2	1	2
11	0	3	2	1	2
12	0	3	2	2	2
13	0	4	2	1	3
14	0	4	2	2	3
15	0	4	3	2	3
16	0	5	3	3	4
17	0	5	3	3	4
18	0	5	3	4	4
19	1	0	1	1	2
20	1	0	1	2	2
21	1	0	2	1	2
22	1	1	1	2	3
23	1	1	2	2	3
24	1	1	2	2	3
25	1	2	2	3	3
26	1	2	3	2	3
27	1	2	3	3	3
28	1	3	4	2	4
29	1	3	4	3	4
30	1	3	4	3	4
31	1	4	5	3	5
32	1	4	5	4	5
33	1	4	6	4	5
34	1	5	6	5	6
35	1	5	7	5	6
36	1	5	7	6	7
Total	–	90	99	82	112

X2 to X5 are quantitative variable

[a]female = 0, male = 1

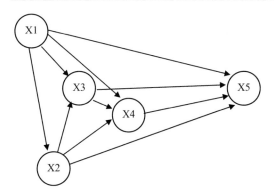

Fig. 2.43   Path diagram of five variables

2.12   Compare the various effects of the independent variables of above causal model.

2.13   Suppose there are causal relationships between five variables as presented in path diagram of Fig. 2.43. Can do you name this variables?

2.14   Given the data of Table 2.10, check the technical assumptions and hypotheses of above path analysis. Is it a fitted path diagram?

2.15   How much is the coefficient of determination of model?

2.16   Complete the effect matrix of above causal model.

2.17   Compare and interpret the various effects of the independent variables of above causal model.

Table 2.10   Distribution of X1 to X5 a probability sample of people

No.	X1	X2	X3	X4	X5
1	0	0	0	0	4
2	0	0	1	1	4
3	0	0	1	1	4
4	0	0	0	0	4
5	0	0	1	1	4
6	0	0	1	1	5
7	0	1	2	2	5
8	0	1	2	1	5
9	0	1	1	1	5
10	0	2	3	2	5
11	0	2	1	1	5
12	0	2	1	1	5

(continued)

Table 2.10   (continued)

No.	X1	X2	X3	X4	X5
13	0	2	3	2	5
14	0	2	2	1	5
15	0	2	2	1	5
16	0	2	2	2	5
17	0	2	2	1	5
18	0	2	2	2	5
19	0	3	2	2	6
20	0	3	2	2	6
21	0	3	2	1	6
22	0	4	3	4	7
23	0	4	3	3	7
24	0	4	3	3	7
25	0	4	3	4	7
26	1	0	1	2	5
27	1	0	2	1	5
28	1	0	1	2	5
29	1	1	2	1	5
30	1	1	2	3	6
31	1	1	2	2	6
32	1	1	2	2	6
33	1	2	3	3	6
34	1	2	2	3	6
35	1	3	2	3	6
36	1	3	4	2	7
37	1	3	4	3	7
38	1	3	4	3	7
39	1	3	4	3	7
40	1	3	4	3	7
41	1	4	4	3	7
42	1	4	5	4	8
43	1	4	5	4	8
44	1	4	6	4	8
45	1	5	6	5	9
46	1	5	6	5	9
47	1	6	7	6	10
48	1	6	7	6	10
Total	23	110	130	113	291

References

Allison PD (1999) Multiple regression: a primer. Sage, Thousand Oaks CA

Darlington RB, Hayes AF (2017b) Regression analysis and linear models, concepts applications and implementation. Guilford Press, New York

de Vaus D (2002) Surveys in social research, 5th ed. Crows Nest NSW, Allen & Unwin

Foster J, Barkus E, Yavorsky C (2006) Understanding and using advanced statistics, Sage, London

Mueller JH, Schuessler KF, Costner HL (1977) Statistical reasoning in sociology, 3rd ed. Houghton Mifflin Company, Boston

Further Reading

The following books contain chapters or sections which deal directly with path analysis or related topics such as causality, causal models, causal inference and causal analysis:

Best H, Wolf C (eds) (2015) The sage handbook of regression analysis and causal inference. Sage, London (For path analysis see Chap. 13)

Blalock HM (ed) (1971) Causal models in the social sciences. Aldine-Atherton, Chicago. This classic work is an excellent collection of essays on the theory and application of causal modeling

Cohen J, Cohen P, West SG, Aiken LS (2003) Applied multiple regression/correlation analysis for the behavioral sciences. Erlbaum, Mahwah, NJ. It is a comprehensive book with emphasis on graphical presentations

Darlington RB, Hayes AF (2017) Regression analysis and linear models, concepts, applications and implementation. Guilford Press, New York (For path analysis see Chap. 15)

Freedman DA (2010) Statistical models and causal inference: a dialogue with the social sciences. Cambridge University Press, New York. It is an overview of the foundations and limitations of statistical modeling

Heise DR (1975) Causal analysis. Wiley, New York. It focuses on linear models in causal analysis (For path analysis see Chap. 4)

Keith TZ (2015) Multiple regression and beyond: an introduction to multiple regression and structural equation modeling. Routledge, New York (For path analysis see Chaps. 11–13)

Morgan SL (ed) (2013) Handbook of causal analysis for social research. Springer, New York. It covers a series of good articles on causal analysis

Shipley B (2004) Cause and correlation in biology: a user's guide to path analysis, structural equations and causal inference. Cambridge University Press, Cambridge, UK. It contains a series of statistical methods for exploring cause-effect relationships between variables

Sloman S (2005) Causal models: how people think about the world. Oxford University Press, New York. It is a good book about causality and its role in everyday life

Tabachnick BG, Fidell LS (2014) Using multivariate statistics, 6th ed. Pearson Education, Edinburgh Gate Harlow (For path analysis see Chap. 14)

Logistic Regression Analysis

3

This chapter describes how a logistic regression can show the effects of a set of independent variables on the probability of occurrence of an event. It shows how the effects of variables are controlled that is necessary to obtain the logistic coefficient. It also illustrates how the pure effect of each of the independent variables can be specified. Finally, it shows how the qualitative variables can be used as binary or dummy variables and what the appropriate logistic regression method for testing hypotheses is.

As the multiple linear regression, the *logistic regression* is a statistical technique for examining the assumed relationships between the independent variables and the dependent variable. The logistic regression analysis is similar to the linear regression analysis, except that the dependent variable in the linear regression is a quantitative (continuous) variable, while in the logistic regression is a binary (dichotomous or binomial) variable. Therefore, the effects of the independent variables on the dependent variable are the effects on the *probability of occurrence* of an *event* which is a given category of the dependent variable.

Here, as the multiple linear regression, first we theoretically assume a set of the independent variables have simultaneously causal impacts on the dependent variable (the probability of occurrence of the event). Thus, we have a set of hypotheses about cause-and-effect relationships between the independent variables and the dependent variable which should be

empirically tested. Then, we collect the appropriate data and use the logistic regression to test our hypotheses.

> The logistic regression is a statistical technique to show the effects of a set of independent variables on the probability of occurrence of an event.

If as we assumed, there are relationships between the independent variables and the dependent variable in reality, the logistic regression can predict and recreate these relationships as an equation. The *logistic regression equation* displays how the probability of occurrence of the event (one category of the dependent variable) is related to each of the independent variables.

Here, we illustrate it with a few simple examples.

Example 3.1 Suppose a researcher based on the argument that the higher the social status, the more the social participation, assumes the participation of men in elections is more than women, because men have higher social status. Then, she or he collected relevant data by interviewing with a probability sample of people in a town (Table 3.1). ◄

In this example with one independent variable (gender), before using the logistic regression equation, we calculate the probability of an event

© The Editor(s) (if applicable) and The Author(s), under exclusive license to Springer Nature Switzerland AG 2020, corrected publication 2021
H. Nayebi, *Advanced Statistics for Testing Assumed Causal Relationships*,
University of Tehran Science and Humanities Series, https://doi.org/10.1007/978-3-030-54754-7_3

Table 3.1 Distribution of variables, Example 3.1

No.	X gender[a]	Y participation[b]
1	0	0
2	0	0
3	0	0
4	0	0
5	0	0
6	0	1
7	0	1
8	0	0
9	0	0
10	0	0
11	0	0
12	0	0
13	0	1
14	0	1
15	0	1
16	0	1
17	0	0
18	0	0
19	0	1
20	0	1
21	1	0
22	1	0
23	1	0
24	1	0
25	1	0
26	1	1
27	1	1
28	1	1
29	1	0
30	1	1
31	1	1
32	1	1
33	1	1
34	1	1
35	1	1
36	1	1
37	1	1
38	1	1
39	1	1

(continued)

Table 3.1 (continued)

No.	X gender[a]	Y participation[b]
40	1	1
41	1	1
42	1	1
43	1	1
44	1	1

[a]Female = 0, male = 1
[b]No = 0, Yes = 1

(participation in election) for different categories of that variable in order to compare them.

Here, given 8 out of 20 women have participated in the election (Table 3.1), the probability of women's participation in election (denoted by P_F) is:

$$P_F = \frac{N_F}{N}$$
$$= \frac{8}{20} \tag{3.1}$$
$$= 0.40$$

where N_F is the number of women and N is total number of cases.

And, given 18 out of 24 men have participated in the election (Table 3.1), the probability of men's participation in election (P_M) is:

$$P_M = \frac{N_M}{N}$$
$$= \frac{18}{24} \tag{3.2}$$
$$= 0.75$$

These results indicate that the probability of men's participation in the election (0.75) is very more than one of women (0.40). In other words gender has effect on the political participation.

In the logistic regression, the probability of the occurrence of an event is differently expressed by a measure, called *odds*. The odds, which are key measure in the logistic regression, are the ratio of the probability of the occurrence of an event (P) to the probability of the non-occurrence of the event $(1 - P)$:

$$odds = \frac{P}{1 - P} \qquad (3.3)$$

In the above example, the odds of women are:

$$odds_F = \frac{P_F}{1 - P_F}$$
$$= \frac{0.40}{1 - 0.40} \qquad (3.4)$$
$$= 0.667$$

This means that the ratio of the probability of participating in election (P) to the probability of not participating ($1 - P$) for women is 0.67.

And, the odds of men are:

$$odds_M = \frac{P_M}{1 - P_M}$$
$$= \frac{0.75}{1 - 0.75} \qquad (3.5)$$
$$= 3.00$$

The odds of men are much more than the odds of women. In other words, the probability of men's participation in election is much more than of women. Indeed, the odds of men are 4.5 times greater than of women:

$$\frac{odds_M}{odds_F} = \frac{3.00}{0.667} \qquad (3.6)$$
$$= 4.55$$

Thus, the odds are a way to present the effect of a variable (or variables) on the probability of the occurrence of an event. In the next section we see how this measure is expressed as a logistic regression equation.

> The odds are the ratio of the probability of an event will occur over the probability of the same event will not occur.

3.1 Logistic Regression Equation

In logistic regression, the odds are presented as an exponential function of a set of independent variables:

$$odds = e^{B_0 + B_1X_1 + B_2X_2 + \cdots + \cdots B_jX_j + \cdots} \qquad (3.7)$$

Or simply:

$$odds = e^{B_0} \times e^{B_1X_1} \times e^{B_2X_2} \times \cdots \times e^{B_jX_j} \times \cdots \qquad (3.8)$$

where e is a mathematical constant, the base of the natural logarithm (approximately equal to 2.718281828459); X_j is jth independent variable; B_j is the logistic coefficient of X_j and B_0 is the constant logistic coefficient.

The logistic regression equation implies that the odds of the event are function of e to the power of a linear combination of the independent variables (Eq. 3.7). In the logistic regression equation, the *logistic coefficient* B of an independent variable X indicates that one unit increases in X changes the odds of the event by a multiple of e to the power of B, holding all other independent variables constant. And, the *logistic constant coefficient* B_0, shows if the values of all independent variables are zero, the odds are equal to e to the power of it.

In above example, the logistic regression equation is as follows:

$$odds = e^{B_0} \times e^{B_1X_1} \qquad (3.9)$$

The running the logistic regression of participation in election (Y) on gender (X) in SPSS results the Variables in the Equation output (Fig. 3.1) which provides the logistic coefficients for a logistic regression equation with one independent variable.[1] The column B of this output includes the regression coefficients for this equation. By replacing these coefficients in the above equation, it will be as follows:

$$odds = e^{-0.405} \times e^{1.504X_1} \qquad (3.10)$$

In this equation, the constant coefficient (B_0) means when X1 (gender) is zero (i.e. cases are women), the odds are $e^{-0.405}$ (or 0.667):

---

[1]Logistic Regression Command in SPSS is described at Sect. 3.6.1.

Variables in the Equation

		B	S.E.	Wald	df	Sig.	Exp(B)
Step 1[a]	Gender	1.504	.656	5.254	1	.022	4.500
	Constant	-.405-	.456	.789	1	.374	.667

a. Variable(s) entered on step 1: Gender.

Fig. 3.1 Variables in the Equation of the logistics regression of participation on gender, Example 3.1

$$
\begin{aligned}
odds &= e^{-0.405} \times e^{1.504(0)} \\
&= e^{-0.405} \times e^{0} \\
&= e^{-0.405} \times 1 \qquad\qquad (3.11) \\
&= e^{-0.405} \\
&= 0.667
\end{aligned}
$$

In other words, the ratio of probability of women's participation in the election to not participation is 0.667, which is presented in the column Exp(B) in Fig. 3.1. It is the same value as the value of Eq. 3.4.

> The logistic coefficient B of an independent variable X indicates that one unit increases in X changes the odds of the event by a multiple of e to the power of B, holding all other independent variables constant.

The logistic coefficient of X1 (gender) suggests when X1 (gender) is 1, the odds change by a multiple of e to the power of 1.504:

$$
\begin{aligned}
odds &= e^{-0.405} \times e^{1.504(1)} \\
&= 0.667 \times 4.500 \qquad\qquad (3.12) \\
&= 3.00
\end{aligned}
$$

This value indicates that the odds of men (the ratio of probability of men's participation in the election to not participation) are 3.00, which are the same value as the value of Eq. 3.5. Also, it means that the odds of men (3.00) are $e^{1.504}$ (or 4.500) times greater than the odds of women.

In this instance, only with one binary independent variable, the results of logistic regression

equation (Eqs. 3.11 and 3.12) may be obtained by the simple ratio of probabilities (Eqs. 3.4 and 3.5). But when there is more than one independent variable, the calculation of the odds requires logistic regression equation.

> The logistic constant coefficient indicates when the values of all other independent variables are zero, the odds are equal to e to the power of its value.

Example 3.2 Suppose as previous example, a researcher based on the same argument that the higher the social status, the more the social participation, assumes the participation of men in elections is more than women. Moreover, she or he assumes that the participation of individuals with higher education is more than of individuals with lower education. Then, she or he collected relevant data by interviewing with a probability sample of people in a town (Table 3.2) in order to test these two hypotheses: (1) the participation of men in elections is more than of women and (2) the participation of individuals with higher education is more than of individuals with low education. ◀

Here, there is a logistic regression model with two independent variables. The result of the logistic regression of participation in election (Y) on gender (X1) and education (X2) with the same coding (female coded as 0 and male as 1; no participation in election coded as 0 and participation coded as 1) is present in Fig. 3.2. It provides the regression logistic coefficients for a logistic regression equation with two independent variables:

Table 3.2 Distribution of variables, Example 3.2

No.	X1 gender[a]	X2 education	Y participation[b]
1	0	0	0
2	0	0	0
3	0	0	0
4	0	0	0
5	0	0	0
6	0	0	0
7	0	0	1
8	0	1	0
9	0	1	0
10	0	1	0
11	0	1	0
12	0	1	0
13	0	1	0
14	0	1	1
15	0	1	1
16	0	1	1
17	0	2	0
18	0	2	0
19	0	2	0
20	0	2	1
21	0	2	1
22	1	0	0
23	1	0	0
24	1	0	0
25	1	0	0
26	1	0	1
27	1	0	1
28	1	0	1
29	1	1	0
30	1	1	0
31	1	1	1
32	1	1	1
33	1	1	1
34	1	1	1
35	1	1	1
36	1	1	1
37	1	1	1
38	1	2	1

(continued)

Table 3.2 (continued)

No.	X1 gender[a]	X2 education	Y participation[b]
39	1	2	1
40	1	2	1
41	1	2	1
42	1	2	1

[a]Female = 0, male = 1
[b]No = 0, Yes = 1

$$odds = e^{B_0} \times e^{B_1 X_1} \times e^{B_2 X_2} \qquad (3.13)$$

By replacing the logistic coefficients (Bs) of Fig. 3.2 in the above equation, the odds of participation in election will be:

$$odds = e^{-2.111} \times e^{2.140 X_1} \times e^{1.155 X_2} \qquad (3.14)$$

The logistic coefficient B_0, the constant coefficient of logistic regression suggests when both variables of X1 (gender) and X2 (education) are zero the odds are equal to e to the power of −2.111:

$$
\begin{aligned}
odds &= e^{-2.111} \times e^{2.140(0)} \times e^{1.155(0)} \\
&= e^{-2.111} \times e^0 \times e^0 \\
&= e^{-2.111} \times 1 \times 1 \qquad (3.15) \\
&= e^{-2.111} \\
&= 0.121
\end{aligned}
$$

This means that the odds of women who are literate are $e^{-2.111}$ or 0.121. In other words, the ratio of probability of illiterate women's participation in the election to not participation is 0.121, which is very low.

The logistic coefficient of X1 (gender) suggests when it is 1 (gender = man) and education is controlled, the odds change by a multiple of e to the power of 2.140:

$$
\begin{aligned}
odds &= e^{-2.111} \times e^{2.140(1)} \times e^{1.155 X_2} \\
&= e^{-2.111} \times 8.501 \times e^{1.155 X_2}
\end{aligned} \qquad (3.16)
$$

It means the probability of participation of men in the election to not participation is $e^{2.140}$ (or 8.501) times greater than of women, by

Variables in the Equation

		B	S.E.	Wald	df	Sig.	Exp(B)
Step 1ª	Gender	2.140	.780	7.523	1	.006	8.501
	Education	1.155	.535	4.659	1	.031	3.173
	Constant	-2.111-	.805	6.874	1	.009	.121

a. Variable(s) entered on step 1: Gender, Education.

Fig. 3.2 Variables in the Equation of the logistics regression of participation in election on gender and education, Example 3.2

Variables in the Equation

		B	S.E.	Wald	df	Sig.	Exp(B)
Step 1ª	gender	.246	.121	4.133	1	.042	1.279
	race	.583	.154	14.436	1	.000	1.792
	education	.058	.017	11.126	1	.001	1.059
	Constant	-.222-	.218	1.033	1	.309	.801

a. Variable(s) entered on step 1: gender, race, education.

Fig. 3.3 Variables in the Equation of logistics regression of participation in election on gender, race and education, Example 3.3

holding education constant. It indicates the gender has effect on the participation in election.

The logistic coefficient of X2 (education) indicates that for one year increase in education, the odds change by a multiple of e to the power of 1.155 (or 3.173), holding gender constant:

$$odds = e^{-2.111} \times e^{2.140X_1} \times e^{1.155(1)}$$
$$= e^{-2.111} \times e^{2.140X_1} \times 3.173 \quad (3.17)$$

It means the probability of participation in the election to probability not participation become $e^{1.155}$ (or 3.173) times greater for one additional year of education by holding gender constant.

This example used a hypothetical data, but the next example uses true data.

Example 3.3 In this example, as previous example based on the same argument that the higher the social status, the more the social participation, we assume the participation of men in elections is more than women, of individuals with higher education is more than of individuals with low education and also of the majority

(white) is higher than the minority (black). Here we test these three hypotheses by using the data of the General Social Survey 1972[2] which is a probability sample of US people. ◄

The result of the logistic regression of participation in election (Y) on gender (X1), race (X2) and education (X3) is present in Fig. 3.3.[3] By replacing the logistic coefficients of Fig. 3.3 in the appropriate logistic equation, the odds of participation in election will be:

$$odds = e^{-0.222} \times e^{0.246X_1} \times e^{0.583X_2} \times e^{0.058X_3}$$
$$(3.18)$$

[2]http://www.gss.norc.org/get-the-data/spss.
[3]In this file, we recoded VOTE68 (Did R vote in 1968 election) into a different variable, named Participation as voted = 1 and not voted = 0, SEX (Respondents sex) into a different variable, named gender as female = 0 and male = 1, RACE (Race of respondent) into a different variable, named race as black = 0 and white = 1. Also, EDUC (Highest year of school completed) named as education.

Here, the constant of logistic regression suggests the odds are e to the power of -0.222 when all independent variables (X1 to X3) are zero. In other words, in 1972, the probability of participating in election to the probability of not participating for individuals who were female, black and illiterate was 0.801.

The logistic coefficient of X1 (gender) suggests when it is 1 (gender = male) and race and education are controlled, the odds change by a multiple of e to the power of 0.246; i.e. the probability of participation of men in the election to not participation is $e^{0.264}$ (or 1.279) times greater than of women by holding race and education constant. In other words, the gender had effect on the participation in election.

The logistic coefficient of X2 (race) suggests the probability of participation of white in the election to not participation was $e^{0.583}$ (or 1.792) times greater than of black by holding gender and education constant. This means that the race had effect on the participation in election.

Finally, the logistic coefficient of X3 (education) indicates that the odds become e to the power of 0.058 (or 1.059) times greater for one year increase in education by holding gender and race constant.

3.1.1 Direction of Logistic Coefficient

When the logistic coefficient (B) of an independent variable (X) is positive, e to the power of it (e^{BX}) is more than 1, and indicates the increase of the probability of occurrence of the event (the odds become greater). In other words, a *positive logistic coefficient* shows that the independent variable makes the odds greater.

Where the logistic coefficient is negative, e to the power of it (e^{BX}) is less than 1, and indicates the decrease of the probability of occurrence of the event (the odds become lower). In other words, a *negative logistic coefficient* shows that the independent variable makes the odds lower.

However, if the logistic coefficient of an independent variable is zero, e to the power of it is equal to one: $e^{(0)X} = e^0 = 1$. In that case, it

doesn't lead to increase or decrease the probability of occurrence of the event; the odds don't change. In other words, the independent variable has not any effect on the dependent variable.

In summary, the logistic regression coefficient greater than 0 increases the odds, the logistic regression coefficient lower than 0 decreases the odds and logistic regression coefficient of 0 leaves the odds unchanged.

> A positive logistic coefficient indicates that the independent variable makes the odds greater.

Thus, when in reality there are causal relationships between the independent variables and a binary dependent variable, the *logistic regression* can recreate these relationships as an equation. In this equation, the coefficient of each independent variable indicates how and in what direction it affects the probability of the occurrence of an event to the probability of the non-occurrence of the event (which is a given category of the dependent variable), while control for the other independent variables.

> A negative logistic coefficient indicates that the independent variable makes the odds lower.

But if there isn't any relationship between an independent variable and the dependent variable, its coefficient will be zero. Where even one of the logistic regression coefficients is zero, the logistic regression model isn't verified and dose not consider as a fitted model.

> The logistic coefficient of zero indicates that the independent variable doesn't change the odds.

3.1.2 Fitted Model

A *logistic regression model* is a theoretical model of a set of independent variables which have simultaneously causal relationships with a binary dependent variable. In other words, the assumed causal relationships are based on a theoretical reasoning. Therefore, maybe it exists in reality, may it dose not. When it exists in reality it is a fitted model, a confirmed model.

A *fitted logistic regression model* is one that: (1) all logistic coefficients of independent variables are nonzero; (2) if data are collected from a probability sample, the statistical significance levels of all logistic coefficients of independent variables are less than 0.05. In that case, the logistic coefficients are generalizable to the population (see significance level in Chap. 1); and (3) all hypotheses are confirmed (observed relationships are according to the theoretical expectations).

For example, the logistic regression model in Example 3.1 (the logistic regression of participation in election on gender) is a fitted logistic regression model, since: (1) its logistic coefficient isn't zero; (2) that coefficient is significant at level less than 0.05 (see the column Sig in Fig. 3.1); and (3) its hypothesis is confirmed (the probability of men's participation in the election to probability not participation is more than of women, and not vice versa). Also, the logistic regression model in examples 3.2 and 3.3 are fitted logistic regression model[4] (why these two model are fitted model?)

Thus, a fitted logistic regression model indicates that there are the assumed relationships between the independent variables and the binary dependent variable in a given population.

> In a fitted logistic regression model, all logistic coefficients of independent variable are nonzero, the statistical significance levels of all coefficients are less than 0.05 and all relations are according to the theoretical expectations.

Goodness-of-Fit

There are several measures for displaying the goodness-of-fit of a fitted logistic regression

model. The *goodness-of-fit measures* are mainly used to compare the fitted logistic regression models; the higher the goodness-of-fit, the better the logistic regression model.

For example, the Cox & Snell R Square (one of the goodness-of-fit measures) of the logistic regression of participation in election on gender in Example 3.1, which is 0.120 (Fig. 3.4) is lower than of the logistic regression of participation in election on gender and education in Example 3.2, which is 0.274 (Fig. 3.5). Thus, the second model with two independent variables (gender and education) is better than the first model with one independent variable (gender).

> The goodness-of-fit measure of a fitted logistic regression model shows that how well it fits the data.

3.2 Controlling for Independent Variables

As the linear regression, the logistic coefficients vary in different subsets of the independent variables, because the calculations of logistic coefficients take into account the effect of the other independent variables in order to maximize the probability of occurrence of the event. Indeed, the *controlling* in logistic regression means that the logistic coefficient for each independent variable is calculated while holding all other independent variables constant.

This controlling is applied in the calculation formula of logistic coefficient. But there is a simple way for understanding the meaning of controlling for the other independent variables. The power of the logistic regression function $(e^{B_0 + B_1X_1 + B_2X_2 + \cdots + B_jX_j + \cdots + B_kX_k})$ can be considered as z:

$$Z = B_0 + B_1X_1 + B_2X_2 + \cdots + B_kX_k$$

which is a linear function of the independent variables of X_1 to X_k. Then, as linear regression we can save the residuals of the linear regression

---

[4]Do you think the regression model of Example 3.3 is a fitted regression model nearly half a century later? You can test your hypotheses by the data of the General Social Survey 2018 (see Exercise 3.8).

Fig. 3.4 Model summary,
Example 3.1

Model Summary

Step	-2 Log likelihood	Cox & Snell R Square	Nagelkerke R Square
1	53.913[a]	.120	.162

a. Estimation terminated at iteration number 4 because parameter estimates changed by less than .001.

Fig. 3.5 Model summary,
Example 3.2

Model Summary

Step	-2 Log likelihood	Cox & Snell R Square	Nagelkerke R Square
1	44.791[a]	.274	.365

a. Estimation terminated at iteration number 5 because parameter estimates changed by less than .001.

of an independent variable on the other independent variables as a new variable and the residual of the linear regression of the dependent variable (Z) on those other independent variables as another new variable. The regression of the residual of the dependent variable on the residual of the independent variable yields logistic coefficient which is the same logistic coefficient of the independent variable in logistic analysis.

For example, if the power of the logistic regression equation of Example 3.2 $(e^{-2.111 + 2.140X_1 + 1.155X_2})$ is considered as z:

$$Z = -2.111 + 2.140X_1 + 1.155X_2$$

We can save the residual of the linear regression of X2 on the X1 as a new variable, ResX2 and the residual of the linear regression of the Z on X1 as another variable, ResZ, as shown in Table 3.3 (for the linear regression, see Chap. 1).

Now, as Fig. 3.6 indicates, the coefficient in the linear regression of the residual of Z (ResZ) on the residual of X2 (ResX2) is 1.115 which is the same logistic coefficient of the X2 in the logistic regression in Example 3.2 (Eq. 3.14).

3.3 Pure Effect

Where there is a fitted logistic regression model, we can specify the pure effect of each of the independent variables on the odds, the ratio of the probability of the occurrence of an event to the probability of the non-occurrence of the event. The *pure effect* in logistic regression is the unique effect of an independent variable on the odds. Indeed, the pure effect is the effect of an independent variable after removing effects of the other independent variables affecting it.

To specify the pure effects of the independent variables on the dependent variable involves following stages:

1. Drawing logistic path diagram
2. Testing logistic path diagram
3. Saving residual variables
4. Standardizing independent variables
5. Obtaining standardized logistic coefficients.

3.3.1 Drawing Logistic Path Diagram.

As the linear regression, in the logistic regression analysis, to specify the unique effect of each of the independent variables on the dependent

Table 3.3 Distributions of variables of Example 3.2 with Z and residuals

No.	X1[a]	X2	Y[b]	Z	ResX2	ResZ
1	0	0	0	−2.111	−0.905	−1.045
2	0	0	0	−2.111	−0.905	−1.045
3	0	0	0	−2.111	−0.905	−1.045
4	0	0	0	−2.111	−0.905	−1.045
5	0	0	0	−2.111	−0.905	−1.045
6	0	0	0	−2.111	−0.905	−1.045
7	0	0	1	−2.111	−0.905	−1.045
8	0	1	0	−0.956	0.095	0.110
9	0	1	0	−0.956	0.095	0.110
10	0	1	0	−0.956	0.095	0.110
11	0	1	0	−0.956	0.095	0.110
12	0	1	0	−0.956	0.095	0.110
13	0	1	0	−0.956	0.095	0.110
14	0	1	1	−0.956	0.095	0.110
15	0	1	1	−0.956	0.095	0.110
16	0	1	1	−0.956	0.095	0.110
17	0	2	0	0.199	1.095	1.265
18	0	2	0	0.199	1.095	1.265
19	0	2	0	0.199	1.095	1.265
20	0	2	1	0.199	1.095	1.265
21	0	2	1	0.199	1.095	1.265
22	1	0	0	0.029	−0.905	−1.045
23	1	0	0	0.029	−0.905	−1.045
24	1	0	0	0.029	−0.905	−1.045
25	1	0	0	0.029	−0.905	−1.045
26	1	0	1	0.029	−0.905	−1.045
27	1	0	1	0.029	−0.905	−1.045
28	1	0	1	0.029	−0.905	−1.045
29	1	1	0	1.184	0.095	0.110
30	1	1	0	1.184	0.095	0.110
31	1	1	1	1.184	0.095	0.110
32	1	1	1	1.184	0.095	0.110
33	1	1	1	1.184	0.095	0.110
34	1	1	1	1.184	0.095	0.110
35	1	1	1	1.184	0.095	0.110
36	1	1	1	1.184	0.095	0.110
37	1	1	1	1.184	0.095	0.110
38	1	2	1	2.339	1.095	1.265
39	1	2	1	2.339	1.095	1.265

(continued)

Table 3.3 (continued)

No.	X1[a]	X2	Y[b]	Z	ResX2	ResZ
40	1	2	1	2.339	1.095	1.265
41	1	2	1	2.339	1.095	1.265
42	1	2	1	2.339	1.095	1.265
Total		38		0.168	-0.010	0.000

[a] Female = 0, male = 1
[b] No = 0, Yes = 1

variable requires to draw a diagram which is a theoretical model that expresses the assumed *causal* relationships between variables of a fitted regression model. However, here we name it "logistic path diagram" in order to distinct it from "path diagram" in the path analysis (Chap. 2), since this theoretical model is only limited to the independent variables.

In a logistic path diagram, each independent variable is connected by an arrow to the other independent variable on which has causal effect. Indeed, every arrow indicates a hypothesis about causal connection between two independent variables. All hypotheses are based on theoretical reasoning which explains the causal relationships between them.

> Pure effect of an independent variable is its unique effect on the dependent variable.

3.3.2 Testing Logistic Path Diagram

Testing a logistic path diagram is to test its hypotheses about causal relationships between the independent variables. In a logistic path diagram, the test of this kind of hypotheses requires as much linear regression as the number of endogenous variables. As mentioned in Chap. 2, an endogenous variable is an independent variable which is affected by the other independent variables. Here if the assumptions of linear regression are not severely violated (see Sect. 1.8) and all of this kind of hypotheses (explicit hypotheses about causal relationships between the independent variables and implicit

Coefficients[a]

Model		Unstandardized Coefficients B	Std. Error	Standardized Coefficients Beta	t	Sig.
1	(Constant)	.000	.000		47644.089	.000
	ResX2	1.155	.000	1.000	150059981.800	.000

a. Dependent Variable: ResZ

Fig. 3.6 Coefficients of the regression of ResZ on ResX2

hypotheses about the lack of causal relationship between them) are confirmed it is a fitted logistic path diagram.

3.3.3 Saving Residual Variables

In a fitted logistic path diagram, to specify the pure effect of an endogenous variable requires to remove the effects of the other independent variables affecting it and to save it as a residual variable. Indeed, a residual variable of an endogenous variable is the residuals of the linear regression of that endogenous variable on the other independent variables affecting it (for details of obtain residuals, see Sect. 1.5, and for Save Residual Command, see Sect. 3.6.2). However, an exogenous variable doesn't require this procedure, because it is an independent variable which isn't affected by other variables.

Now, all exogenous variables and residual variables are uncorrelated variables. As mentioned in Chap. 1, uncorrelated independent variables are variables that have not relationship with each other.

> An endogenous variable is an independent variable that is affected by the other independent variables in logistic path diagram.

3.3.4 Standardizing Independent Variables

In this stage all these uncorrelated independent variables should be standardized in order to can

compare their effects. However, the dependent variable doesn't require to be standardized, since the result of the logistic regression of a binary dependent variable on standardized independent variables is the same as the result of the logistic regression of a standardized binary dependent variable.[5]

> An exogenous variable is an independent variable that isn't affected by any independent variable in path diagram.

3.3.5 Obtaining Standardized Logistic Coefficients

The logistic regression of the dependent variable on these uncorrelated standardized independent variables yields the standardized logistic coefficients. The *standardized logistic coefficient* of such an uncorrelated variable indicates that the odds of an event uniquely become e to the power of it times greater (or smaller) when that variable increase by one standard deviation. Thus, the comparable pure effects of the independent variables on the odds of the event are provided; the standardized logistic coefficients of uncorrelated independent variables are their pure effects on the odds of an event.

---

[5]For standardizing a variable, each of its values is converted to standard measure which is also called standard score or z-score. A standard measure is the difference between a true value and the mean divided by standard deviation. SPSS automatically standardizes variables (for Standardize Command in SPSS, see Sect. 3.6.3).

Here, the stages of specifying the pure effects of the independent variables on the dependent variable in a fitted logistic regression model are presented by using the following examples.

> The standardized logistic coefficients of a set of uncorrelated variables are their pure effects on the odds of an event.

Example 3.4 As previous examples, a researcher based on the argument that the higher the social status, the more the social participation, assumes three variables of gender (X1), education (X2) and occupational prestige (X3) are three major factors which have causal effects on the participation in election (Y), because men, who have higher education and who occupy higher prestigious occupations have higher social status. ◀

Moreover, since in contemporary societies the social status of men is usually higher than women, then the education and occupational prestige of men are averagely higher than women; i.e. the gender has causal effect on education and occupational prestige. Also, the education is one of the main factors of occupational prestige; i.e. the education has causal effect on obtaining high level occupations.

Now, suppose the data of Table 3.4 were collected by interviewing with a probability sample of employed people in a town in order to test these hypotheses.

As Fig. 3.7 shows, the result of the logistic regression of participation in election (Y) on gender (X1), education (X2) and occupational prestige (X3) indicate that all logistic coefficients are nonzero and all of them are significant at level less than 0.05. Also, its hypotheses are confirmed (the participation of men, white and who with higher occupational prestige is more). Therefore, this model is a fitted logistic regression model. Therefore, we can specify the pure effects of the independent variables.

Drawing Logistic Path Diagram

Logistic path diagram presented in Fig. 3.8 shows the mentioned hypotheses about causal relationships between the independent variables of Example 3.4. Here, the hypotheses of this logistic path diagram about the independent variables are: (1) X1 has causal effect on X2; (2) X1 has causal effects on X3; and (3) X2 has causal effects on X3.

Testing Logistic Path Diagram

Here, the test of the logistic path diagram with two endogenous variables (X2 and X3) requires two linear regression: (1) the linear regression of X2 on X1, which shows that there is a significant relationship between these two variables (Fig. 3.9); and (2) the linear regression of X3 on X1 and X2, which indicates that X3 has significant relationships with both variables (Fig. 3.10). Moreover, all observed relations are according to the theoretical expectations, since are positive which mean that men have higher education (first hypothesis); and men and who with higher education have higher occupational (hypotheses 2 and 3). Also, suppose the assumptions of linear regression are not severely violated in these linear regressions. Thus, all hypotheses of logistic path diagram about the independent variables are confirmed. So, we can conclude that the logistic path diagram of Example 3.4 fits the data. In other words, this logistic path diagram (Fig. 3.8) is a fitted one.

Saving Residual Variables

The fitted logistic path diagram of Fig. 3.8 shows that endogenous variable X2 is affected by X1. So, we remove the effect of X1 from it and save it as a residual variable (ResX2). Also, endogenous variable X3 is affected by X1 and X2. So, we remove the effects of X1 and X2 from it and save it as a residual variable (ResX3)[6] (see Table 3.5).

Standardizing

In this stage, we save the exogenous X1 and two residual variables (ResX2 and ResX3) as standardized variables ZX1, ZResX2 and ZResX3 (see Table 3.5).

---

[6]1 Save Residual Command in SPSS is described at Sect. 3.6.2.

Table 3.4 Distributions of variables, Example 3.4

No.	X1[a]	X2	X3	Y[b]	No.	X1	X2	X3	Y
1	0	0	0	0	38	1	2	2	1
2	0	0	0	0	39	1	3	1	1
3	0	0	0	1	40	1	3	2	1
4	0	0	0	1	41	1	3	2	1
5	0	1	1	0	42	1	3	2	1
6	0	1	1	0	43	1	3	2	1
7	0	1	1	0	44	1	3	2	1
8	0	1	1	0	45	1	3	2	1
9	0	1	1	0	46	1	3	2	1
10	0	1	1	0	47	1	3	2	1
11	0	1	1	0	48	1	3	2	1
12	0	1	1	0	49	1	3	3	1
13	0	1	1	0	50	1	3	3	1
14	0	1	1	0	51	1	3	3	1
15	0	1	1	0	52	1	3	3	1
16	0	1	1	0	53	1	3	3	1
17	0	2	1	0	54	1	3	3	1
18	0	2	1	0	55	1	3	3	1
19	0	2	2	0	56	1	3	3	1
20	0	2	2	0	57	1	4	1	1
21	0	2	2	1	58	1	4	2	1
22	0	2	2	1	59	1	4	2	1
23	0	3	0	0	60	1	4	3	1
24	0	3	0	0	61	1	4	3	1
25	0	3	3	1	62	1	4	3	1
26	0	3	3	1	63	1	4	4	1
27	0	4	1	0	64	1	4	4	1
28	0	4	1	0	65	1	4	4	1
29	0	4	1	1	66	1	4	4	1
30	0	4	1	1	67	1	4	4	1
31	1	1	1	0	68	1	4	4	1
32	1	1	1	0	69	1	5	4	1
33	1	2	2	0	70	1	5	4	1
34	1	2	2	0	71	1	5	5	1
35	1	2	2	1	72	1	5	5	1
36	1	2	2	1	Total		188	145	
37	1	2	2	1					

[a]Female = 0, male = 1
[b]No = 0, Yes = 1

Variables in the Equation

		B	S.E.	Wald	df	Sig.	Exp(B)
Step 1[a]	Gender	1.643	.829	3.927	1	.048	5.170
	Education	.741	.357	4.305	1	.038	2.099
	Occupational prestige	1.121	.562	3.982	1	.046	3.068
	Constant	-3.878-	1.100	12.425	1	.000	.021

a. Variable(s) entered on step 1: Gender, Education, Occupational prestige.

Fig. 3.7 Variables in the equation of the logistic regression of Y on X1, X2 and X3, Example 3.4

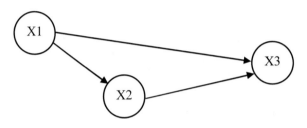

Fig. 3.8 Logistic path diagram, Example 3.4

Coefficients[a]

Model		Unstandardized Coefficients		Standardized Coefficients	t	Sig.
		B	Std. Error	Beta		
1	(Constant)	1.733	.202		8.591	.000
	X1 Gender	1.505	.264	.563	5.696	.000

a. Dependent Variable: X2 Education

Fig. 3.9 Coefficients of the linear regression of X2 on X1, Example 3.4

Coefficients[a]

Model		Unstandardized Coefficients		Standardized Coefficients	t	Sig.
		B	Std. Error	Beta		
1	(Constant)	.223	.206		1.083	.283
	X1 Gender	.891	.227	.357	3.923	.000
	X2 Education	.487	.085	.522	5.732	.000

a. Dependent Variable: X3 Occupational prestige

Fig. 3.10 Coefficients of the linear regression of X3 on X1 and X2, Example 3.4

Table 3.5 Distributions of variables, Example 3.4 with residual standardized variables

No.	X1	X2	X3	Y	ResX2	ResX3	ZX	ZResX2	ZResX3
1	0	0	0	0	-1.733-	-.223-	-1.175-	-1.580-	-.287-
2	0	0	0	0	-1.733-	-.223-	-1.175-	-1.580-	-.287-
3	0	0	0	1	-1.733-	-.223-	-1.175-	-1.580-	-.287-
4	0	0	0	1	-1.733-	-.223-	-1.175-	-1.580-	-.287-
5	0	1	1	0	-.733-	.290	-1.175-	-.668-	.375
6	0	1	1	0	-.733-	.290	-1.175-	-.668-	.375
7	0	1	1	0	-.733-	.290	-1.175-	-.668-	.375
8	0	1	1	0	-.733-	.290	-1.175-	-.668-	.375
9	0	1	1	0	-.733-	.290	-1.175-	-.668-	.375
10	0	1	1	0	-.733-	.290	-1.175-	-.668-	.375
11	0	1	1	0	-.733-	.290	-1.175-	-.668-	.375
12	0	1	1	0	-.733-	.290	-1.175-	-.668-	.375
13	0	1	1	0	-.733-	.290	-1.175-	-.668-	.375
14	0	1	1	0	-.733-	.290	-1.175-	-.668-	.375
15	0	1	1	0	-.733-	.290	-1.175-	-.668-	.375
16	0	1	1	0	-.733-	.290	-1.175-	-.668-	.375
17	0	2	1	0	.267	-.197-	-1.175-	.243	-.254-
18	0	2	1	0	.267	-.197-	-1.175-	.243	-.254-
19	0	2	2	0	.267	.803	-1.175-	.243	1.038
20	0	2	2	0	.267	.803	-1.175-	.243	1.038
21	0	2	2	1	.267	.803	-1.175-	.243	1.038
22	0	2	2	1	.267	.803	-1.175-	.243	1.038
23	0	3	0	0	1.267	-1.683-	-1.175-	1.154	-2.174-
24	0	3	0	0	1.267	-1.683-	-1.175-	1.154	-2.174-
25	0	3	3	1	1.267	1.317	-1.175-	1.154	1.700
26	0	3	3	1	1.267	1.317	-1.175-	1.154	1.700
27	0	4	1	0	2.267	-1.170-	-1.175-	2.066	-1.512-
28	0	4	1	0	2.267	-1.170-	-1.175-	2.066	-1.512-
29	0	4	1	1	2.267	-1.170-	-1.175-	2.066	-1.512-
30	0	4	1	1	2.267	-1.170-	-1.175-	2.066	-1.512-
…	…	…	…	…	………	………	..……..	……...	..……..
…	…	…	…	…	………	………	..……..	……...	..……..
…	…	…	…	…	………	………	..……..	……...	..……..
69	1	5	4	1	1.762	.452	.839	1.606	.583
70	1	5	4	1	1.762	.452	.839	1.606	.583
71	1	5	5	1	1.762	1.452	.839	1.606	1.875
72	1	5	5	1	1.762	1.452	.839	1.606	1.875
Total		188	145	46	0.000	0.000	0.000	0.000	0.000

Variables in the Equation

		B	S.E.	Wald	df	Sig.	Exp(B)
Step 1[a]	ZX1	2.273	.508	20.035	1	.000	9.711
	ZResX2	1.412	.451	9.807	1	.002	4.106
	ZResX3	.868	.435	3.982	1	.046	2.382
	Constant	1.274	.456	7.816	1	.005	3.576

a. Variable(s) entered on step 1: ZX1, ZResX2, ZResX3.

Fig. 3.11 Variables in the equation of the logistic regression of Y on ZX1, ZResX2 and ZResX3, Example 3.4

Obtaining Standardized Logistic Coefficients
Finally, the logistic regression of the dependent variable Y on these uncorrelated standardized independent variables ZX1, ZResX2 and ZResX3 results the standardized logistic coefficients of these three variables (Fig. 3.11). These standardized logistic coefficients are the main components of their comparable pure effect on the odds of participation in election.

Here, the logistic standardized coefficient of X1 indicates that one standard unit increase in X1 (gender) makes the odds of the participation in election become $e^{2.273}$ (or 9.711) times greater. Also, the logistic standardized coefficient of ZResX2 shows that after removing the effect of X1 (gender) from X2 (education), one standard unit increase in it makes the odds of the participation in election become $e^{1.412}$ (or 4.106) times greater. And, the logistic standardized coefficient of ZResX3 indicates that after removing the effects of X1 (gender) and X2 (education) from occupational prestige (X3) one standard unit increase in it makes the odds of the participation in election become $e^{0.868}$ (or 2.382) times greater.

In this way, it is overt that X1 has largest pure effect on Y and X3 has lowest one; the pure effect of X1 is over two times more than the pure effect of X2 ($9.711/4.106 = 2.365$) and over four times more than the pure effect of X3 ($9.711/2.382 = 4.077$). Also, the pure effect of X2 is over one and half times more than the pure effect of X3 ($4.106/2.382 = 1.736$).

Here, we used a hypothetical data, but the next example is based on true data.

Example 3.5 Suppose the higher social position, the more satisfaction of social life such as the satisfaction with financial situation. As a result, individuals who are white, male, with higher occupational prestige and higher income are more satisfaction with financial situation. In other words, race (X1), gender (X2), occupational prestige (X3) and income (X4) have causal effects on financial satisfaction (X5). ◀

Also, based on the same argument, individuals who are white and male and occupy more prestigious occupations have more income: race (X1), gender (X2), and occupational prestige (X3) have causal effects on income (X4).

Moreover, who are male and white have higher occupational prestige: race (X1) and gender (X2) have causal effects on occupational prestige (X3).

However, it is obvious there isn't relationship between race (X1) and gender (X2). Also, at least in Western countries, the difference between women and men has been very small in many respects such as assess to high level occupations. So, we don't assume any relationship between gender (X2) and occupational prestige (X3).

Here, we test this causal model by using the data of the General Social Survey 2018[7] which is a probability sample of US people.[8] The result of

[7]http://www.gss.norc.org/get-the-data/spss.
[8]In this file, we recoded RACE (Race of respondent) as race (X1) coded as black and other = 0 and white = 1; SEX (Respondents sex) as gender (X2) coded as female = 0 and male = 1; PRESTIGE (R's occupational prestige score 1970) as occupational prestige (X3), REALINC (Family income in constant $) as income (X4) which has rounded to thousand dollars (e.g.

Variables in the Equation

		B	S.E.	Wald	df	Sig.	Exp(B)
Step 1ª	Race	.306	.121	6.350	1	.012	1.358
	Gender	.196	.101	3.731	1	.053	1.216
	Occupational prestige	.009	.004	5.127	1	.024	1.009
	Family income	.021	.002	150.954	1	.000	1.022
	Constant	-2.289-	.200	130.487	1	.000	.101

a. Variable(s) entered on step 1: Race, Gender, Occupational prestige, Family income .

Fig. 3.12 Variables in the equation of the logistic regression of X5 on X1 to X4, Example 3.5

Fig. 3.13 Logistic path diagram, Example 3.5

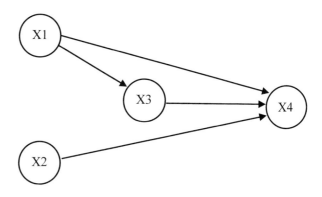

the logistic regression of X5 on X1 to X4 indicates that this model is a fitted logistic regression model, since all logistic coefficients are nonzero and all of them are significant at level less than 0.05 (Fig. 3.12). Moreover, all logistic coefficients are positive which means they are according to the theoretical expectations: who are white, male and have higher occupational prestige and more income have more satisfaction of their financial situation. Thus, all hypotheses are confirmed. Therefore, this model is a fitted logistic regression model and we can specify the pure effects of the independent variables.

Now, we draw a logistic path diagram for the assumed relationships between independent variables of this fitted model.

Drawing logistic path diagram
Logistic path diagram presented in Fig. 3.13 shows the mentioned hypotheses about causal relationships between the independent variables. Here, the explicit hypotheses of this logistic path diagram about the independent variables are: (1) X1 has causal effect on X3; (2) X1 has causal effect on X4, too; (3) X2 has causal effect on X4; and (4) X3 has causal effect on X4. Moreover, the implicit hypotheses of this logistic path diagram about the independent variables are: (1) X2 has not any relationship with X1; and (2) X2 has not any relationship with X3, too.

Testing Logistic Path Diagram
Here, the test of the logistic path diagram with two endogenous variables (X4 and X3) requires two linear regression: (1) the linear regression of X4 on X1 to X3, which shows that the relationships between X4 with those three variables are significant (Fig. 3.14); and (2) the linear regression of X3 on X1, which indicates that X3 has significant relationships with X1 (Fig. 3.15). Finally, as Fig. 3.16 shows, X2 has not any significant relationship with X1 and X3.

---

665 dollar to 1 thousand dollar, or 2345 to 2 thousand dollars, because the difference by one dollar has not significant effect). Also, we recoded SATFIN (Satisfaction with financial situation) as a binary variable, 1 = Satisfied and 0 = Not satisfied (sum of "more or less" and "not at all"), named financial satisfaction (X5).

Coefficients[a]

Model		Unstandardized Coefficients		Standardized Coefficients		
		B	Std. Error	Beta	t	Sig.
1	(Constant)	-6.271-	2.399		-2.614-	.009
	X1 Race	8.990	1.459	.128	6.161	.000
	X2 Gender	3.771	1.282	.060	2.941	.003
	X3 Occupational prestige	.714	.047	.312	15.082	.000

a. Dependent Variable: X4 Family income

Fig. 3.14 Coefficients of the linear regression of X4 on X1 to X3, Example 3.5

Coefficients[a]

Model		Unstandardized Coefficients		Standardized Coefficients		
		B	Std. Error	Beta	t	Sig.
1	(Constant)	41.692	.549		76.005	.000
	X1 Race	4.098	.642	.133	6.382	.000

a. Dependent Variable: X3 Occupational prestige

Fig. 3.15 Coefficients of the linear regression of X3 on X1, Example 3.5

Correlations

		X2 Gender	X1 Race	X3 Occupational prestige
X2 Gender	Pearson Correlation	1	.020	.001
	Sig. (2-tailed)		.333	.954
	N	2348	2348	2248
X1 Race	Pearson Correlation	.020	1	.133[**]
	Sig. (2-tailed)	.333		.000
	N	2348	2348	2248
X3 Occupational prestige	Pearson Correlation	.001	.133[**]	1
	Sig. (2-tailed)	.954	.000	
	N	2248	2248	2248

**. Correlation is significant at the 0.01 level (2-tailed).

Fig. 3.16 Correlations of X2 with X1 and X3, Example 3.5

Variables in the Equation

		B	S.E.	Wald	df	Sig.	Exp(B)
Step 1[a]	ZX1	.268	.054	24.690	1	.000	1.307
	ZX2	.137	.050	7.439	1	.006	1.147
	ZResX3	.326	.051	41.071	1	.000	1.385
	ZResX4	.621	.051	150.954	1	.000	1.860
	Constant	-.865-	.051	283.372	1	.000	.421

a. Variable(s) entered on step 1: ZX1, ZX2, ZResX3, ZResX4.

Fig. 3.17 Variables in the equation of the logistic regression of X5 on ZX1, ZX2, ZResX3 and ZResX4, Example 3.5

Moreover, all observed relations are according to the theoretical expectations, since are positive which mean that white, men, and who with higher occupational prestige have more income; and white have higher occupational prestige. Also, suppose the assumptions of linear regression are not severely violated in two linear regressions.

Thus, all (explicit and implicit) hypotheses of the logistic path diagram about the independent variables are confirmed. So, we can conclude that the logistic path diagram of Example 3.5 fits the data. In other words, the logistic path diagram is a fitted one.

Saving Residual Variables

As the fitted logistic path diagram of Fig. 3.13 shows the endogenous variable X3 is affected by X1. So, we remove the effect of X1 from it and save it as a residual variable ResX3. Also, the endogenous variable X4 is affected by X1 to X3. So, we remove the effects of X1 to X3 from it and save it as a residual variable ResX4.

Standardizing Variables

Now, we save the exogenous X1 and X2 and two residual variables ResX3 and ResX4 as standardized variables ZX1, ZX2, ZResX3 and ZResX4.

Obtaining Standardized Logistic Coefficients

Finally, the logistic regression of the dependent variable X5 on these uncorrelated standardized independent variables ZX1, ZX2, ZResX3 and ZResX4 results the logistic standardized logistic coefficients of these variables (Fig. 3.17).

The coefficient of each of these variables indicates that one standard unit increase in a variable makes the odds of participation in election become e to the power of that coefficient times greater.

In this way, it is overt that income (X4) has largest pure effect (1.860) on the financial satisfaction (X5). The occupational prestige (X3) has large pure effect (1.385), too. But gender (X2) has the lowest pure effect among these variables.

3.4 Qualitative Variables

As in the linear regression, the quantitative variables can be used as binary or dummy variables in the logistic regression.

3.4.1 Binary Variable

The *binary variable* is a qualitative variable with two categories. The logistic coefficient of a binary variable indicates that the ratio of the odds of a category to the odds of other category is e to power of it, holding all other independent variables constant.

The categories of a binary variable are usually coded as 0 and 1. Although, to give code 0 or 1 to this or that category doesn't matter, it is better to give code 1 to a category which makes the odds rise. In that case, the logistic coefficient of a binary independent variable indicates the value by which the odds of the occurrence of the event raise in the category with code 1 compared with other category by controlling the effect of the other independent variables.

A multi- categories variable which is a qualitative variable with more than two categories may be turned into binary variable and used in logistic regression analysis. In this way, it involves a category which is to be compared with a new category which includes all of other categories.

3.4.2 Dummy Variable

The *qualitative variable* with more than two categories can be turned into a set of *dummy variables*, too. In turning a variable into the dummy variables, one of categories will be set aside for the *base of comparison* (a category with the lowest probability of occurrence is preferred). Then each of other categories is turned into a binary (0 and 1) variable. A dummy variable indicates the presence of a given category of the original variable (coded 1) and the absence of other categories (coded 0). The logistic coefficient of a dummy variable represents the value by which the odds of the event change compared with the base category, holding other independent variables constant.

The logistic regression command can automatically construct the dummy variables (Logistic Dummy Command is described at the end of chapter, Sect. 3.6.4).

> The logistic coefficient of a binary variable shows that the ratio of the odds of its categories is e to power of it.

Example 3.6 Suppose a researcher based on the same argument of Example 3.2 that the higher the social status, the more the social participation, assumes in addition to gender and education, the ethnicity of people effects the participation in election, too: the participation of men, who with higher education and the majority ethnic group are greater, because their social status are higher. ◄

Also, suppose the data for testing these hypotheses are collected from the survey of a probability sample of people in a town (Table 3.6). Here, gender is coded as female = 0 and male = 1, the ethnicity as ethnic A (the majority ethnic group) = 1, ethnic B = 2 and ethnic C = 3. The dependent variable, participation in election is coded as yes = 1 and no = 0.

The logistic regression of Y (participation in election) on X1 (gender), X2 (education) and X3 (ethnicity) with specification X3 as categorical covariates (see Logistic Dummy Command, Sect. 3.6.4) yields a Categorical Variables Codings table which indicated how the dummy variables are created.

Figure 3.18 indicates that Ethnicity (X3) has turned into two dummy variables: Ethnicity (1) with code 1 for everybody who is a member of ethic A and code 0 for everybody who isn't its member; Ethnicity (2) with code 1 for everybody who is a member of ethnic B, and code 0 for everybody who isn't its member. The third category (ethnic C) has not turned into a dummy variable and remained as a base for comparison.

Interpretation of Coefficient of Dummy Variables

The logistic regression coefficient (B) of a dummy variable indicates that its odds are e^B times greater (or lower) compared with the base category by controlling effects of the other independent variables. Thus, in Example 3.6, the logistic coefficient of the Ethnicity (1) indicates that its odds are $e^{2.249}$ times greater than the ethnic C while controlling for effects of gender and education (Fig. 3.19). In other words, the probability of participation in election to the probability of not participation for the members of ethnic A is $e^{2.249}$ (or 9.480) times greater than one of the members of ethnic C. And, the logistic coefficient of the Ethnicity (2) indicates that its odds are $e^{1.580}$ times greater than the ethnic C while controlling for effects of gender and education (Fig. 3.19). In other words, the probability of participation in election to the probability of not participation for the members of ethnic B are $e^{1.580}$ (or 4.855) times greater than one of the members of ethnic C.

Table 3.6 Distributions of variables of Example 3.6

No.	X1	X2	X3	Y	No.	X1	X2	X3	Y
1	0	0	1	1	46	0	0	2	0
2	1	0	1	1	47	0	0	2	0
3	0	1	1	0	48	0	0	2	0
4	0	1	1	0	49	0	0	2	0
5	0	1	1	0	50	0	0	2	0
6	0	1	1	0	51	1	0	2	1
7	0	1	1	1	52	1	0	2	1
8	0	1	1	1	53	1	0	2	1
9	0	1	1	1	54	1	0	2	1
10	0	1	1	1	55	1	0	2	1
11	0	1	1	1	56	0	1	2	1
12	0	1	1	1	57	0	1	2	1
13	0	1	1	1	58	0	1	2	1
14	0	1	1	1	59	0	1	2	1
15	1	1	1	1	60	0	1	2	1
16	1	1	1	1	61	1	1	2	0
17	1	1	1	1	62	1	1	2	0
18	1	1	1	1	63	1	1	2	1
19	1	1	1	1	64	1	1	2	1
20	1	1	1	1	65	1	1	2	1
21	1	1	1	1	66	1	1	2	1
22	1	1	1	1	67	1	2	2	0
23	0	2	1	0	68	1	2	2	1
24	0	2	1	0	69	0	0	3	0
25	0	2	1	1	70	0	0	3	0
26	0	2	1	1	71	0	0	3	0
27	0	2	1	1	72	0	0	3	0
28	0	2	1	1	73	0	0	3	0
29	0	2	1	1	74	0	0	3	0
30	0	2	1	1	75	0	0	3	0
31	0	2	1	1	76	0	0	3	0
32	0	2	1	1	77	1	0	3	0
33	1	2	1	1	78	1	0	3	0
34	1	2	1	1	79	1	0	3	0
35	1	2	1	1	80	1	0	3	0
36	1	2	1	1	81	1	0	3	0
37	1	2	1	1	82	1	0	3	0
38	1	2	1	1	83	1	0	3	0
39	1	2	1	1	84	1	0	3	0

(continued)

Table 3.6 (continued)

No.	X1	X2	X3	Y	No.	X1	X2	X3	Y
40	1	2	1	1	85	0	1	3	1
41	1	2	1	1	86	0	1	3	1
42	1	2	1	1	87	1	1	3	1
43	0	0	2	0	88	1	1	3	1
44	0	0	2	0	Total		79	154	
45	0	0	2	0					

X1 gender: female = 0, male = 1; X2 education by year; X3 ethnicity: ethnic A = 1, ethnic B = 2, ethnic C = 3; Y participation in election: yes = 1, no = 0

> The logistic coefficient of a dummy variable shows the value by which the odds of the event change compared with the base category.

3.5 Logistic Regression Method

As well as linear regression, the logistic regression can be analyzed by different methods. However, the usual method is the enter method in which all independent variables enter into analysis at once. This method is used primarily when the researcher assumes a set of independent variables affect the dependent variable, i.e. researcher has a theoretical model.

However, there are other methods which are used to explore statistical relations between a set of variables and the dependent variable which isn't focus of this book.

> The enter method is the appropriate logistic regression method for testing hypotheses.

3.6 Commands

This section includes commands for running the Logistic Regression Command, Save Residual Command, Standardization Command, and Logistic Dummy Command.

Categorical Variables Codings

| | | Frequency | Parameter coding | |
			(1)	(2)
X3 Ethnicity	1 ethnic A	42	1.000	.000
	2 ethnic B	26	.000	1.000
	3 ethnic C	20	.000	.000

Fig. 3.18 Categorical variables codings, Example 3.6

Variables in the Equation

		B	S.E.	Wald	df	Sig.	Exp(B)
Step 1[a]	Gender	1.491	.630	5.609	1	.018	4.444
	Education	1.327	.549	5.854	1	.016	3.770
	Ethnicity			6.337	2	.042	
	Ethnicity(1)	2.249	.931	5.834	1	.016	9.480
	Ethnicity(2)	1.580	.776	4.150	1	.042	4.855
	Constant	-2.630-	.766	11.776	1	.001	.072

a. Variable(s) entered on step 1: Gender, Education, Ethnicity.

Fig. 3.19 Variables in the equation table, Example 3.6

3.6.1 Logistic Regression Command

To carry out a logistic regression analysis in SPSS:

1. In the Data Editor (data file) as shown in Fig. 3.20, select **Analyze → Regression → Binary Logistic …**,
2. In the **Logistic Regression** dialogue box (Fig. 3.21), select the dependent variable from the list of variables and transfer it into the **Dependent:** box and the independent variables into the **Covariates:** box,
3. Click on the **OK** button.

3.6.2 Save Residual Command

To save the residual of an endogenous variable after removing the effects of the other independent variables affecting it as a residual variable:

1. In the Data Editor (data file) as shown in Fig. 3.22, select **Analyze → Regression → Linear…**.
2. In the **Linear Regression** dialogue box (Fig. 3.23), select the endogenous variable from the list of variables and transfer it into the **Dependent:** box and the other independent variables which affect it into the **Independent(s):** box.
3. Click on the **Save…** button.
4. In the **Linear Regression: Save** dialogue box (Fig. 3.24), in the **Residuals** section, click on **Unstandardized**.
5. Click on the **Continue** button.
6. In the **Linear Regression** dialogue box, click **OK**.

3.6.3 Standardize Command

To standardize a variable in data file of SPSS:

Fig. 3.20 Data editor, analyze, regression, binary logistic …

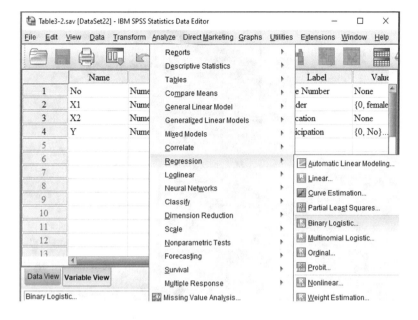

Fig. 3.21 Logistic regression dialogue box

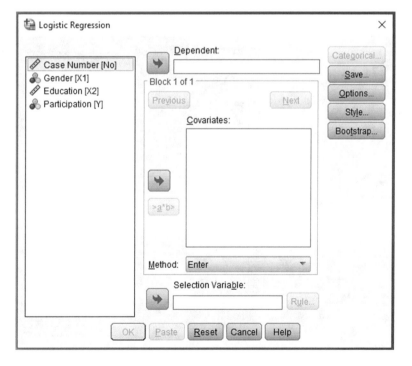

Fig. 3.22 Data editor, analyze, regression, linear…

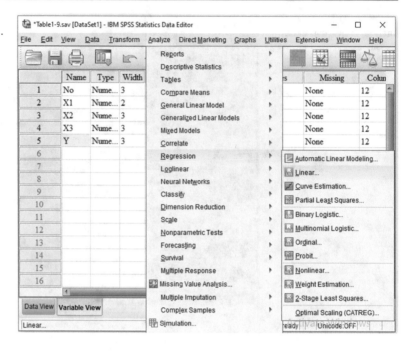

Fig. 3.23 Linear regression dialog box

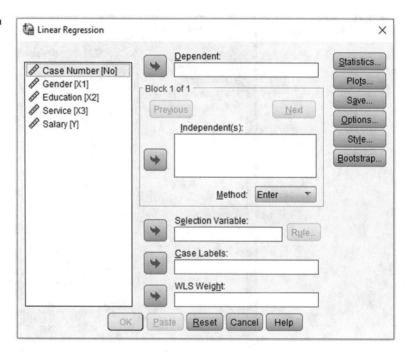

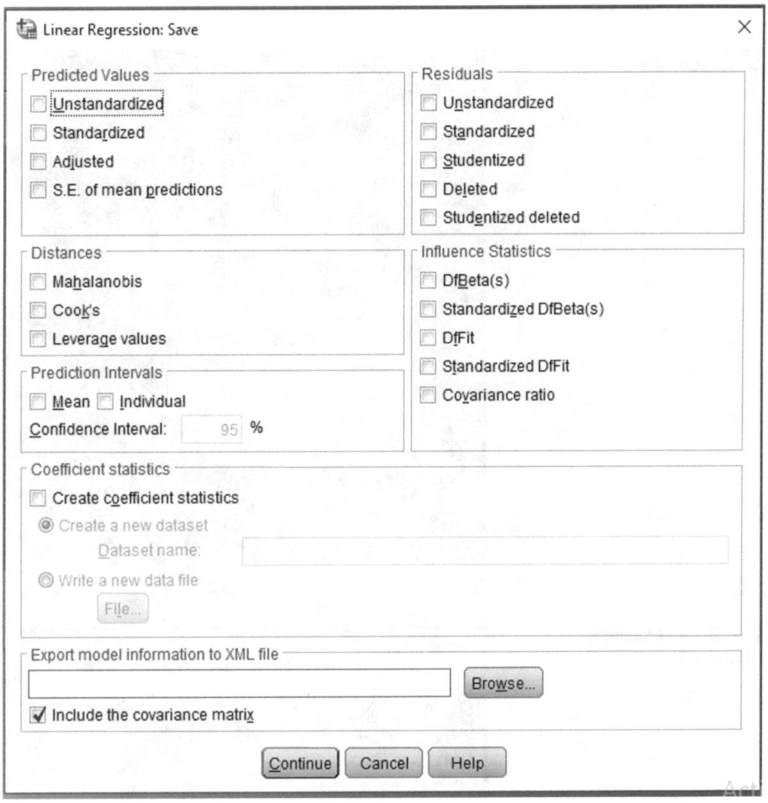

Fig. 3.24 Linear regression: save dialogue box

1. In the Data Editor (data file) as shown in Fig. 3.25, select **Analyze → Descriptive Statistics → Descriptives…**,
2. In the **Descriptives** dialogue box (Fig. 3.26) select the variable(s) from the list of left box and transfer it into the **Variable(s):** box,
3. Click on the **Save standardized values as variables**,
4. Click on the **OK** button.

3.6.4 Logistic Dummy Command

To make dummy variables:
1. In the **Logistic Regression** dialogue box (Fig. 3.21) after selecting the dependent variable from the list of variables and transfer it into the **Dependent:** box and the independent variables into the **Covariates:** box, click on the **Categorical…** button.

2. In the **Logistic Regression: Define Categorical Variables** dialogue box (Fig. 3.27), select the independent variable(s) which is to turn into dummy variable from the **Covariates:** box and transfer it into the **Categorical Covariates:** box.
3. Click on the **Continue** button.
4. In the **Logistic Regression** dialogue box, click **OK**.

In this Command, the default option for the base category (reference category) is **Last** category. You can change it and make the first category as the base category. In that case, in step 2:

In the **Logistic Regression: Define Categorical Variables** dialogue box (Fig. 3.27), click on **First** option and then on **Change** button.

Fig. 3.25 Data editor, descriptive statistics, descriptives…

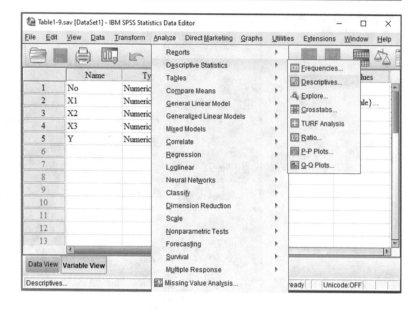

Fig. 3.26 Descriptives dialogue box

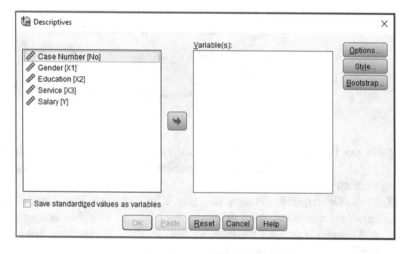

3.7 Summary

This chapter describes the logistic regression. It is a powerful statistical technique for examining the assumed causal relationships between a set of independent variables on the probability of occurrence of an event which is one category of a binary dependent variable. Here, the effects of the variables are presented through odds, which are the ratio of the probability of an event will occur over the probability of the same event will not occur. It is shown how the odds are presented as a function

of a set of independent variables. The assumed causal relationships are confirmed when all logistic coefficients of independent variables are non-zero, all observed relationships are according to the theoretical expectations, and where the data are collected from a probability sample, all logistic coefficients are significant. It is presented how the effects of variables can be controlled that is necessary to obtain the logistic coefficient. It is shown the pure effect of an independent variable is its unique effect on the dependent variable after removing effects of the other independent variables affecting it. To specify the pure effects of the

Fig. 3.27 Logistic regression: define categorical variables dialogue box

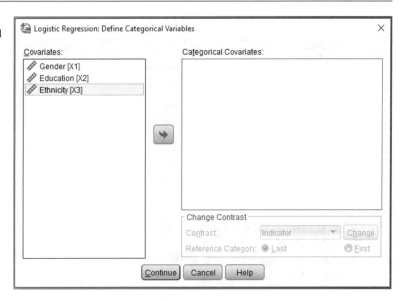

independent variables involves drawing a logistic path diagram, testing logistic path diagram, saving residual variables, standardizing, and obtaining standardized logistic coefficients. In a logistic regression, the qualitative variables can be used as binary or dummy variables. The chapter concludes with showing the appropriate logistic regression method for testing hypotheses.

3.8 Exercises

3.1 A researcher assumed the two independent variables X1 (a binomial variables) and X2 affect the dependent variable Y which is a binary variable. The data for testing this model were collected from an interview with a random sample of individuals in a city (Table 3.7). Are above hypotheses confirmed? Why?

3.2 Is the above logistic regression model a fitted one? Why?

3.3 Determine the logistic regression equation of the above model.

3.4 Interpret the logistic regression coefficients of that equation.

3.5 Is the logistic regression of Y on X1 a fitted model? Why?

3.6 What about the logistic regression of Y on X2?

Table 3.7 Distribution of X1, X2 and Y, Exercise. 3.1

No.	X1	X2	Y
1	0	0	0
2	0	0	0
3	0	2	1
4	0	0	0
5	0	0	0
6	0	1	0
7	0	1	0
8	0	3	1
9	0	1	0
10	0	1	0
11	0	1	1
12	0	1	0
13	0	1	1
14	0	2	0
15	0	2	0
16	0	2	0
17	0	2	1
18	1	0	1
19	0	1	1
20	1	0	0
21	0	1	0
22	1	2	1
23	1	1	1
24	1	1	1

(continued)

Table 3.7 (continued)

No.	X1	X2	Y
25	1	1	1
26	0	2	1
27	1	1	1
28	0	3	1
29	0	3	1
30	1	2	1
Total	8	38	16

3.7 Which of above logistic models is the fittest one? Why?

3.8 Test the hypotheses of Example 3.3 by using the data of the General Social Survey 2018[9] which is a probability sample of US people. Is the logistic regression of occupational prestige on gender, race and education a fitted logistic regression model?[10] Why?

3.9 Refer to the Example 3.4, write the equation of the logistic regression of the participation in election (Y) on gender (X1), education (X2) and occupational prestige (X3).

3.10 Interpret the logistic regression coefficients of above equation.

3.11 A researcher argued four independent variables X1 to X3 have causal effects on X4; X1 and X2 have causal effects on X3; and X1 has causal effect on X2.

(A) Can you name these variables?

(B) Refer to the data of Table 3.8 (collected from a probability sample of people in a town), is the logistic regression of X4 on X1 to X3 a fitted model? Why?

(C) If above model is a fitted one, specify and interpret the pure effect of each of the independent variables, and then compare them.

[9]http://www.gss.norc.org/get-the-data/spss.
[10]In this file, the appropriate variables are: SEX (Respondents sex), RACE (Race of respondent), EDUC (Highest year of school) and VOTE12 (Did R vote in 2012 election).

Table 3.8 Distribution of X1 to X4, Exercise. 3.11

No.	X1	X2	X3	X4
1	0	0	0	0
2	0	0	1	0
3	0	1	1	0
4	0	1	1	0
5	0	0	1	0
6	0	1	1	0
7	0	1	1	0
8	0	1	1	0
9	0	0	1	0
10	0	0	0	0
11	1	1	1	1
12	0	2	1	1
13	0	2	2	1
14	1	2	2	1
15	1	3	2	1
16	1	1	2	1
17	0	2	2	1
18	1	0	2	1
19	1	0	2	1
20	1	2	2	1
21	1	2	3	1
22	1	3	3	1
23	1	2	3	1
24	1	3	3	1
25	0	0	0	0
26	0	0	1	0
27	0	0	1	0
28	0	1	1	0
29	0	2	1	0
30	0	1	1	0
31	0	1	1	0
32	0	0	2	0
33	1	1	2	0
34	0	0	1	0
35	0	1	2	0
36	1	1	1	1
37	0	2	1	1
38	0	2	2	1
39	0	2	2	1

(continued)

Table 3.8 (continued)

No.	X1	X2	X3	X4
40	1	3	2	1
41	0	0	2	1
42	0	0	2	1
43	0	2	2	1
44	1	0	2	1
45	1	2	2	1
46	1	2	2	1
47	1	1	3	1
48	1	2	3	1
49	1	3	3	1
50	1	3	3	1
Total	21	62	83	29

3.12 Suppose, the higher social position, the more happiness. As a result, we can assume individuals who are white, with higher occupational prestige and higher income are more happiness. In other words, race, occupational prestige and income have causal effects on happiness. Also, based on the same argument, individuals who are white and occupy higher occupational prestige have more income: race and occupational prestige have causal effects on income. Moreover, white have higher occupational prestige: race has causal effect on occupational prestige. Now, test these three hypotheses by using the data of the General Social Survey 1972[11] which is a probability sample of US people.[12]

[12]In this file, recode RACE (Race of respondent) as race (X1) coded as Black = 0 and White = 1; consider PRESTIGE (R's occupational prestige score 1970) as occupational prestige (X2), and REALINC (Family income in constant $) as income (X3) and round it to thousand dollars (e.g. 564 dollar to 1 thousand dollar, or 2442 to 2 thousand dollars, because the difference by one dollar has not significant effect), and finally recode HAPPY (General happiness) as a binary variable, 1 = Very Happy and 0 = Not Very Happy (sum of "Petty Happy" and "Not Too Happy"), named happiness (X4).

(A) Is the logistic regression of X4 on X1 to X3 a fitted model? Why?

(B) If above model is a fitted one, specify and interpret the pure effect of each of the independent variables, and then compare them.

3.13 Do you think above logistic regression model is a fitted model nearly half a century later?. You can test your hypotheses by the data of the General Social Survey 2018.[13] In this file, the appropriate variables are: RACE (Race of respondent), PRESTG10 (R's occupational prestige score), REALINC (Family income in constant $) and HAPPY (General happiness).

3.14 Refer to the data of Table 3.9, turn race (X2) into a binary variable and run the logistic regression of environmental attitude (Y) on the above new created independent variable and gender (X1).

(A) Is it a fitted logistic regression model?

(B) Write the equation for above regression model

(C) Interpret the regression coefficients.

3.15 Again, refer to the data of Table 3.9, turn race (X2) into a set of dummy variables. How many dummy variables did you create? Why?

3.16 Run the logistic regression of independent variable Y on the above dummy variables and X1.

(A) Is it a fitted regression model?

(B) Write the equation for above regression model

(C) Interpret the regression coefficients.

3.17 Which of above two models is a fitted regression model?

Table 3.9 Distribution of X1 to X3

No.	X1	X2	X3
1	0	1	0
2	1	1	1
3	0	1	0
4	0	1	1
5	0	1	1
6	0	1	1
7	0	1	1
8	0	1	1
9	0	1	1
10	0	1	1
11	1	1	1
12	1	1	1
13	1	1	1
14	0	1	0
15	0	1	0
16	0	1	1
17	0	1	1
18	0	1	1
19	0	1	1
20	0	1	0
21	1	1	1
22	1	1	1
23	1	1	1
24	1	1	1
25	1	1	1
26	0	2	0
27	0	2	0
28	0	2	0
29	0	2	0
30	1	2	1
31	1	2	1
32	1	2	1
33	0	2	1
34	0	2	1
35	0	2	1
36	0	2	1
37	1	2	0

(continued)

Table 3.9 (continued)

No.	X1	X2	X3
38	1	2	1
39	1	2	1
40	1	2	0
41	1	2	1
42	0	3	0
43	0	3	0
44	0	3	0
45	0	3	0
46	1	3	0
47	1	3	0
48	1	3	1
Total	20	78	31

$X1$ gender: female = 0, male = 1; $X2$ ethnicity: ethnic A = 1, ethnic B = 2, ethnic C = 3; Y environmental attitude: 1 = agree with greenhouse gas reduction policy, 0 = not agree

Further Reading

Agresti A, Finlay B (2009) Statistical methods for the social sciences, 4th ed. Upper Saddle River, New Jersey, Prentice Hall

Best H, Wolf C (eds) (2015) The Sage handbook of regression analysis and causal inference. Sage, London

Chatterjee S, Simonoff JS (2013) Handbook of regression analysis. Wiley, Hoboken, NJ

Cohen J, Cohen P, West SG, Aiken LS (2003) Applied multiple regression/correlation analysis for the behavioral sciences. Erlbaum, Mahwah, NJ

Darlington RB, Hayes AF (2017) Regression analysis and linear models, concepts, applications and implementation. Guilford Press, New York

Eliason SR (1993) Maximum likelihood estimation: logic and practice. Sage, Newbury Park, CA

Harrell FE (2015) Regression modeling strategies with applications to linear models logistic and ordinal regression and survival analysis, 2nd ed. Springer, New York, NY

Harrell FE (2015b) Regression modeling strategies with applications to linear models logistic and ordinal regression and survival analysis, 2nd ed. Springer, New York, NY

Hilbe JM (2009) Logistic regression models. Chapman and Hall/CRC, Boca Raton FL

Hosmer DW, Lemeshow S (2000) Applied logistic regression, 2nd ed. Wiley, New York

Long JS (1997) Regression models for categorical and limited dependent variables. Sage, Thousand Oaks

Menard S (2001) Applied logistic regression analysis, 2nd ed. Thousand Oaks, CA, Sage

Ryan TP (1997) Modern regression methods. Wiley, New York

Tabachnick BG, Fidell LS (2014) Using multivariate statistics, 6th ed. Pearson Education, Edinburgh Gate Harlow

Correction to: Advanced Statistics for Testing Assumed Causal Relationships

Hooshang Nayebi

Correction to:
H. Nayebi, *Advanced Statistics for Testing Assumed Causal Relationships*,
University of Tehran Science and Humanities Series,
https://doi.org/10.1007/978-3-030-54754-7

In the original version of the book, the following correction has been incorporated: The word "casual" has been replaced with "causal" throughout the book, including the cover.

The book has been updated with the changes.

The updated version of the book can be found at
https://doi.org/10.1007/978-3-030-54754-7

H. Nayebi, *Advanced Statistics for Testing Assumed Causal Relationships*,
University of Tehran Science and Humanities Series, https://doi.org/10.1007/978-3-030-54754-7_4

Index

Printed in the United States
by Baker & Taylor Publisher Services